The Reenchantment of the World

The Reenchantment
of the World

MORRIS BERMAN

Cornell University Press

ITHACA AND LONDON

Cornell University Press gratefully acknowledges a grant from the Andrew W. Mellon Foundation that aided in bringing this book to publication.

First published 1981 by Cornell University Press.
First printing, Cornell Paperbacks, 1981.
Published in the United Kingdom by Cornell University Press Ltd.,
Ely House, 37 Dover Street, London W1X 4HQ.

Acknowledgment is made to:
    The Bobbs-Merrill Company, Inc., for excerpts from René Descartes, *Discourse on Method*, translated by Laurence J. Lafleur; copyright © 1950, 1956, by the Liberal Arts Press, Inc.; reprinted by permission of the Liberal Arts Press Division of the Bobbs-Merrill Company, Inc.
    Doubleday & Company, Inc., for permission to quote excerpts from *The Poetry and Prose of William Blake*, edited by David V. Erdman; copyright © 1965 by David V. Erdman and Harold Bloom; and an excerpt from *The Birth of Tragedy and the Genealogy of Morals* by Friedrich Nietzsche, translated by Francis Golffing; copyright © 1956 by Doubleday & Company, Inc.
    Harper & Row, Publishers, Inc., for permission to quote specified brief excerpts from *Steps to an Ecology of Mind* by Gregory Bateson (T. Y. Crowell); copyright © 1972 by Harper & Row, Publishers, Inc.
    Charles Scribner's Sons for permission to use an illustration by Fons van Woerkom from *The Tender Carnivore and the Sacred Game* by Paul Shepard, illustrated by Fons van Woerkom; text copyright © 1973 by Paul Shepard, illustrations copyright © 1973 by Fons van Woerkom.

International Standard Book Number (cloth) 0-8014-1347-0
International Standard Book Number (paper) 0-8014-9225-4
Library of Congress Catalog Card Number 81-67178
Printed in the United States of America
Librarians: Library of Congress cataloging information appears on the last page of the book.

For three friends:
Michael Crisp
David Kubrin
John Trotter

God and philosophy could not live together peacefully; can philosophy survive without God? Once its adversary has disappeared, metaphysics ceases to be the science of sciences and becomes logic, psychology, anthropology, history, economics, linguistics. What was once the great realm of philosophy has today become the ever-shrinking territory not yet explored by the experimental sciences. If we are to believe the logicians, all that remains of metaphysics is no more than the nonscientific residuum of thought—a few errors of language. Perhaps tomorrow's metaphysics, should man feel a need to think metaphysically, will begin as a critique of science, just as in classical antiquity it began as a critique of the gods. This metaphysics would ask itself the same questions as in classical philosophy, but the starting point of the interrogation would not be the traditional one *before* all science but one *after* the sciences.

—Octavio Paz, *Alternating Current*

Contents

Illustrations

Illustrations

Acknowledgments

Several people read all or part of the manuscript version of this book and offered significant criticisms and suggestions, and I am particularly grateful to Paul Ryan, Carolyn Merchant of the University of California, Berkeley, Frederick Ferré of Dickinson College, and W. David Lewis of Auburn University. There is, of course, no unanimous agreement on the final content of the work, and as is usually the case, errors of fact or interpretation are strictly my own. There are also a number of other friends who, although they did not read the manuscript, exerted an important influence on my life through the example of their own, making it possible for me to clarify certain issues that ultimately came to be reflected in this book. Bill Williams, Jack London, David Kubrin, and Deirdre Rand have, over the years, profoundly touched me and even altered my definition of reality, and it is a pleasure for me to acknowledge my debt to them at this time.

I doubt there is any way I can adequately thank my critic and dear friend Michael Crisp, who acted as an astute and untiring reader and who significantly influenced my thinking, particularly in the case of Chapter 3, much of which grew out of discussions we had on the magical tradition. On more than one occasion, Mr. Crisp helped me to resolve some problem of logic or exposition, and I can only hope that his inclusion in the dedication to my book will repay him in some small measure for his great interest and generous assistance.

I wish, finally, to acknowledge my very large debt to John Ackerman at Cornell University Press, whose ruthless editing did much to improve the final form of my manuscript.

This book draws on, and often explicates, the work of Carl Jung,

Acknowledgments

Wilhelm Reich, and Gregory Bateson, among others, but I am not aware of having followed the conceptual framework of any particular philosophical school. It is, nevertheless, a product of its times and reflects a holistic world view that is very much "in the air." Although I have not read all or even most of their work, my outlook has much in common with such writers as R. D. Laing, Theodore Roszak, and Philip Slater, and in many ways we seem to inhabit the same mental universe. In particular, their hope for a humanized culture in which science would play a very different role than it has hitherto is my hope as well.

M. B.

San Francisco

Grateful acknowledgment is also extended to the following for permission to reprint copyright material:

Robert Bly for permission to reprint Poem Number 16 of *The Kabir Book*, version by Robert Bly, published by Beacon Press; copyright © 1971 by Robert Bly.

Cambridge University Press for two diagrams from *Patterns of Discovery* by Norwood Russell Hanson; copyright © 1958, 1965 by Cambridge University Press.

Coward, McCann & Geoghegan, Inc., for two diagrams reprinted by permission of Coward, McCann & Geoghegan, Inc., from *Depression and the Body* by Alexander Lowen, MD; copyright © 1972 by Alexander Lowen, MD.

Grove Press, Inc., for excerpt from *The Labyrinth of Solitude* by Octavio Paz; reprinted by permission of Grove Press, Inc.; copyright © 1961 by Grove Press, Inc.

Harcourt Brace Jovanovich, Inc., and Faber and Faber Ltd., for lines from "Little Gidding" by T. S. Eliot, from *Four Quartets* in T. S. Eliot, *The Complete Poems and Plays, 1909–1950*; copyright © 1952 by Harcourt Brace Jovanovich, Inc.; and in *Collected Poems, 1909–1962*; copyright © 1963 by Faber and Faber Ltd.

Humanities Press, New Jersey, for an excerpt from *The Metaphysical Foundations of Modern Science* by E. A. Burtt; copyright © 1932 by Doubleday & Company, Inc.

Maclen Music, Inc., for lyrics from "When I'm Sixty Four" (John Lennon and Paul McCartney); copyright © 1967 by Northern Songs Limited. All rights for the USA, Mexico and the Philippines controlled by Maclen Music, Inc. Used by permission. All rights reserved.

W. W. Norton & Company, Inc., for lines from Shakespeare's *Henry IV, Part I*, edited by James L. Sanderson; copyright © 1969 by W. W. Norton & Company, Inc.

Oxford University Press for an excerpt from *Micrographia* by Robert Hooke, in vol. XIII of *Early Science in Oxford*, edited by R. T. Gunther.

Penguin Books Ltd., for an excerpt from *The Politics of Experience and the Bird of Paradise* by R. D. Laing; reprinted by permission of Penguin Books Ltd.; copyright © 1967 by R. D. Laing.

Random House, Inc., for two diagrams and specified excerpts from *The Divided Self* by R. D. Laing; copyright © 1962 by Pantheon Books Inc., a Division of Random House, Inc.; and Associated Book Publishers Ltd.; copyright © Tavistock Publications (1959) Ltd. 1960.

Viking Penguin Inc. for permission to quote excerpts from *The World Turned Upside Down* by Christopher Hill; copyright © 1972 by Christopher Hill; and an excerpt from *Alternating Current* by Octavio Paz; copyright © 1973 by Octavio Paz.

Ziff-Davis Publishing Co. for excerpt from Claude Lévi-Strauss interview; reprinted from *Psychology Today* magazine; copyright © 1972, Ziff-Davis Publishing Co.

Introduction:
The Modern Landscape

> You see all round you people engaged in making others live lives
> which are not their own, while they themselves care nothing for
> their own real lives—men who hate life though they fear death.
>
> —William Morris, *News from Nowhere* (1891)

For several years now I have intended to write a semipopular book, dealing with certain contemporary problems, and based on my knowledge of the history of science. In an earlier work, a very technical monograph, I was able only to hint at some of the problems that characterize life in the Western industrial nations, problems that I find profoundly disturbing.[1] I began that study in the belief that the roots of our dilemma were social and economic in nature; by the time I had completed it, I was convinced that I had omitted a whole epistemological dimension. I began to feel, in other words, that something was wrong with our entire world view. Western life seems to be drifting toward increasing entropy, economic and technological chaos, ecological disaster, and ultimately, psychic dismemberment and disintegration; and I have come to doubt that sociology and economics can by themselves generate an adequate explanation for such a state of affairs.

The present book, then, is an attempt to take my previous analysis one step further; to grasp the modern era, from the sixteenth century to the present, as a whole, and to come to terms with the metaphysical presuppositions that define this period. This is not to treat mind, or consciousness, as an independent entity, cut off from material life; I hardly believe such is the case. For purposes of discussion, however, it is often necessary to separate these two aspects of human experience; and although I shall make every effort to demonstrate their interpenetration, my primary focus in this book is the transformations of the human mind. This emphasis stems from my conviction that the fundamental issues confronted by any civilization in its history, or by any person in his or her life, are issues of *meaning*. And historically, our loss of meaning in an ultimate philosophical or religious sense—the split between fact and value which characterizes the modern age—is rooted in the Scientific Revolution of the sixteenth and seventeenth centuries. Why should this be so?

The view of nature which predominated in the West down to the eve of the Scientific Revolution was that of an enchanted world. Rocks, trees, rivers, and clouds were all seen as wondrous, alive, and human beings felt at home in this environment. The cosmos, in short, was a place of *belonging*. A member of this cosmos was not an alienated observer of it but a direct participant in its drama. His personal destiny was bound up with its destiny, and this relationship gave meaning to his life. This type of consciousness—what I shall refer to in this book as "participating consciousness"—involves merger, or identification, with one's surroundings, and bespeaks a psychic wholeness that has long since passed from the scene. Alchemy, as it turns out, was the last great coherent expression of participating consciousness in the West.

The story of the modern epoch, at least on the level of mind, is one of progressive disenchantment. From the sixteenth century on, mind has been progressively expunged from the phenomenal world. At least in theory, the reference points for all scientific explanation are matter and motion—what historians of science refer to as the "mechanical philosophy." Developments that have thrown this world view into question—quantum mechanics, for example, or certain types of contemporary ecological research—have not made any significant dent in the dominant mode of thinking. That mode can best be described as disenchantment, nonparticipation, for it insists on a rigid distinction between observer and

16

observed. Scientific consciousness is alienated consciousness: there is no ecstatic merger with nature, but rather total separation from it. Subject and object are always seen in opposition to each other. I am not my experiences, and thus not really a part of the world around me. The logical end point of this world view is a feeling of total reification: everything is an object, alien, not-me; and I am ultimately an object too, an alienated "thing" in a world of other, equally meaningless things. This world is not of my own making; the cosmos cares nothing for me, and I do not really feel a sense of belonging to it. What I feel, in fact, is a sickness in the soul.

Translated into everyday life, what does this disenchantment mean? It means that the modern landscape has become a scenario of "mass administration and blatant violence,"[2] a state of affairs now clearly perceived by the man in the street. The alienation and futility that characterized the perceptions of a handful of intellectuals at the beginning of the century have come to characterize the consciousness of the common man at its end. Jobs are stupefying, relationships vapid and transient, the arena of politics absurd. In the vacuum created by the collapse of traditional values, we have hysterical evangelical revivals, mass conversions to the Church of the Reverend Moon, and a general retreat into the oblivion provided by drugs, television, and tranquilizers. We also have a desperate search for therapy, by now a national obsession, as millions of Americans try to reconstruct their lives amidst a pervasive feeling of anomie and cultural disintegration. An age in which depression is a norm is a grim one indeed.

Perhaps nothing is more symptomatic of this general malaise than the inability of the industrial economies to provide meaningful work. Some years ago, Herbert Marcuse described the blue- and white-collar classes in America as "one-dimensional." "When technics becomes the universal form of material production," he wrote, "it circumscribes an entire culture; it projects a historical totality—a 'world.'" One cannot speak of alienation as such, he went on, because there is no longer a self to be alienated. We have all been bought off, we all sold out to the System long ago and now identify with it completely. "People recognize themselves in their commodities," Marcuse concluded; they have become what they own.[3]

Marcuse's is a plausible thesis. We all know the next-door neighbor who is out there every Sunday, lovingly washing his car with an ardor that is almost sexual. Yet the actual data on the day-to-day life of the middle and working classes tend to refute Marcuse's notion that for these people, self and commodities have merged, producing what he terms the "Happy Consciousness." To take only two examples, Studs Terkel's interviews with hundreds of Americans, drawn from all walks of life, revealed how hollow and meaningless they saw their own vocations. Dragging themselves to work, pushing themselves through the daily tedium of typing, filing, collecting insurance premiums, parking cars, interviewing welfare applicants, and largely fantasizing on the job—these people, says Terkel, are no longer characters out of Charles Dickens, but out of Samuel Beckett.[4] The second study, by Sennett and Cobb, found that Marcuse's notion of the mindless consumer was totally in error. The worker is not buying goods because he identifies with the American Way of Life, but because he has enormous anxiety about his self, which he feels possessions might assuage. Consumerism is paradoxically seen as a way *out* of a system that has damaged him and that he secretly despises; it is a way of trying to keep *free* from the emotional grip of this system.[5]

But keeping free from the System is not a viable option. As technological and bureaucratic modes of thought permeate the deepest recesses of our minds, the preservation of psychic space has become almost impossible.[6] "High-potential candidates" for management positions in American corporations customarily undergo a type of finishing-school education that teaches them how to communicate persuasively, facilitate social interaction, read body language, and so on. This mental framework is then imported into the sphere of personal and sexual relations. One thus learns, for example, how to discard friends who may prove to be career obstacles and to acquire new acquaintances who will assist in one's advancement. The employee's wife is also evaluated as an asset or liability in terms of her diplomatic skills. And for most males in the industrial nations, the sex act itself has literally become a project, a matter of carrying out the proper techniques so as to achieve the prescribed goal and thus win the desired approval. Pleasure and intimacy are seen almost as a hindrance to the act. But once the ethos of technique and management has permeated the spheres of sexuality and friendship, there is literally no place left to hide. The "widespread climate of anxiety and neurosis" in which we are immersed is thus inevitable.[7]

These details of the inner psychological landscape lay bare the

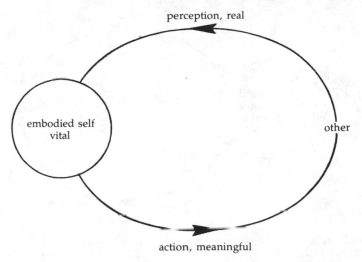

perception, real

embodied self
vital

other

action, meaningful

Figure 1. R. D. Laing's schematic drawing of healthy interaction (from Laing, *The Divided Self,* p. 81).

workings of the System most completely. In a study that purported to be about schizophrenia, but that was for the most part a profile of the psychopathology of everyday life, R. D. Laing showed how the psyche splits, creating false selves, in an attempt to protect itself from all this manipulation.[8] If we were asked to characterize our usual relations with other persons, we might (as a first guess) describe them as pictured in Figure 1 (see above). Here we have self and other in direct interaction, engaging each other in an immediate way. As a result, perception is real, action is meaningful, and the self feels embodied, vital (enchanted). But as the discussion above clearly indicates, such direct interaction almost never takes place. We are "whole" to almost no one, least of all ourselves. Instead we move in a world of social roles, interaction rituals, and elaborate game-playing that forces us to try to protect the self by developing what Laing calls a "false-self system."

In Figure 2, the self has split in two, the "inner" self retreating from the interaction and leaving the body—now perceived as false, or dead (disenchanted)—to deal with the other in a way that is pure theater, while the "inner" self looks on like a scientific observer. Perception is thus unreal, and action correspondingly futile. As Laing points out, we retreat into fantasies at work—and in "love"—and establish a false self (identified with the body and its

19

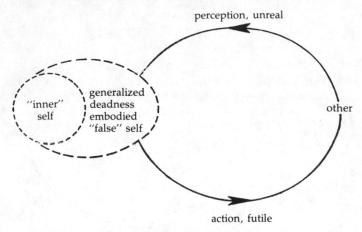

perception, unreal

"inner" self

generalized deadness embodied "false" self

other

action, futile

Figure 2. R. D. Laing's schematic drawing of schizoid interaction (from *The Divided Self,* p. 81).

mechanical actions) which performs the rituals necessary for us to succeed in our tasks. This process begins sometime during the third year of life, is reinforced in kindergarten and grammar school, continues on into the dreary reality of high school, and finally becomes the daily fare of working life.[9] Everyone, says Laing—executives, physicians, waiters, or whatever—playacts, manipulates, in order to avoid being manipulated himself. The aim is the protection of the self, but since that self is in fact cut off from any meaningful intercourse, it suffocates. The environment becomes increasingly unreal as human beings distance themselves from the events of their own lives. As this process accelerates, the self begins to fight back, to nag itself (and thus create a further split) about the existential guilt it has come to feel. We are haunted by our phoniness, our playacting, our flight from trying to become what we truly are or could be. As the guilt mounts, we silence the nagging voice with drugs, alcohol, spectator sports—anything to avoid facing the reality of the situation. When the self-mystification we practice, or the effect of the pills, wears off, we are left with the terror of our own betrayal, and the emptiness of our manipulated "successes."

The statistics that reflect this condition in America alone are so grim as to defy comprehension. There is now a significant suicide rate among the seven-to-ten age group, and teenage suicides tripled between 1966 and 1976 to roughly thirty per day. More than half the patients in American mental hospitals are under twenty-one. In 1977, a survey of nine- to eleven-year-olds on the West Coast found that nearly half the children were regular users of alcohol, and that huge numbers in this age group regularly came to school drunk. Dr. Darold Treffert, of Wisconsin's Mental Health Institute, observed that millions of children and young adults are now plagued by "a gnawing emptiness or meaninglessness expressed not as a fear of what may happen to them, but rather as a fear that nothing will happen to them." Official figures from government reports released during 1971–72 recorded that the United States has 4 million schizophrenics, 4 million seriously disturbed children, 9 million alcoholics, and 10 million people suffering from severely disabling depression. In the early 1970s, it was reported that 25 million adults were using Valium; by 1980, Food and Drug Administration figures indicated that Americans were downing benzodiazepines (the class of tranquilizers which includes Valium) at a rate of 5 billion pills a year. Hundreds of thousands of the nation's children, according to *The Myth of the Hyperactive Child* by Peter Schrag and Diane Divoky (1975), are being drugged in the schools, and one-fourth of the American female population in the thirty-to-sixty age group uses psychoactive prescription drugs on a regular basis. Articles in popular magazines such as *Cosmopolitan* urge sufferers from depression to drop in to the local mental hospital for drugs or shock treatments, so that they can return to their jobs as quickly as possible. "The drug and the mental hospital," writes one political scientist, "have become the indispensable lubricating oil and reservicing factory needed to prevent the complete breakdown of the human engine."[10]

These figures are American in degree, but not in kind. Poland and Russia are world leaders in the consumption of hard liquor; the suicide rate in France has been growing steadily; in West Germany, the suicide rate doubled between 1966 and 1976.[11] The insanity of Los Angeles and Pittsburgh is archetypal, and the "misery index" has been climbing in Leningrad, Stockholm, Milan, Frankfurt and other cities since midcentury. If America is the frontier of the Great Collapse, the other industrial nations are not far behind.

It is an argument of this book that we are *not* witnessing a peculiar twist in the fortunes of postwar Europe and America, an aberration that can be tied to such late twentieth-century problems as inflation, loss of empire, and the like. Rather, we are witnessing the inevitable outcome of a logic that is already centuries old, and

21

which is being played out in our own lifetime. I am not trying to argue that science is the cause of our predicament; causality is a type of historical explanation which I find singularly unconvincing. What I am arguing is that the scientific world view is *integral* to modernity, mass society, and the situation described above. It is *our* consciousness, in the Western industrial nations—uniquely so—and it is intimately bound up with the emergence of our way of life from the Renaissance to the present. Science, and our way of life, have been mutually reinforcing, and it is for this reason that the scientific world view has come under serious scrutiny at the same time that the industrial nations are beginning to show signs of severe strain, if not actual disintegration.

From this perspective, the transformations I shall be discussing, and the solutions I dimly perceive, are epochal, and this is all the more reason not to relegate them to the realm of theoretical abstraction. Indeed, I shall argue that such fundamental transformations impinge upon the details of our daily lives far more directly than the things we may think to be most urgent: this Presidential candidate, that piece of pressing legislation, and so on. There have been other periods in human history when the accelerated pace of transformation has had such an impact on individual lives, the Renaissance being the most recent example prior to the present. During such periods, the meaning of individual lives begins to surface as a disturbing problem, and people become preoccupied with the meaning of meaning itself. It appears a necessary concomitant of this preoccupation that such periods are characterized by a sharp increase in the incidence of madness, or more precisely, of what is seen to define madness.[12] For value systems hold us (*all* of us, not merely "intellectuals") together, and when these systems start to crumble, so do the individuals who live by them. The last sudden upsurge in depression and psychosis (or "melancholia," as these states of mind were then called) occurred in the sixteenth and seventeenth centuries, during which time it became increasingly difficult to maintain notions of salvation and God's interest in human affairs. The situation was ultimately stabilized by the emergence of the new mental framework of capitalism, and the new definition of reality based on the scientific mode of experiment, quantification, and technical mastery. The problem is that this whole constellation of factors—technological manipulation of the environment, capital accumulation based on it, notions of secular salvation that fueled it and were fueled by it—has apparently

22

run its course. In particular, the modern scientific paradigm has become as difficult to maintain in the late twentieth century as was the religious paradigm in the seventeenth. The collapse of capitalism, the general dysfunction of institutions, the revulsion against ecological spoilation, the increasing inability of the scientific world view to explain the things that really matter, the loss of interest in work, and the statistical rise in depression, anxiety, and outright psychosis are all of a piece. As in the seventeenth century, we are again destabilized, cast adrift, floating. We have, as Dante wrote in the *Divine Comedy*, awoken to find ourselves in a dark woods.

What will serve to stabilize things today is fairly obscure; but it is a major premise of this book that because disenchantment is intrinsic to the scientific world view, the modern epoch contained, from its inception, an inherent instability that severely limited its ability to sustain itself for more than a few centuries. For more than 99 percent of human history, the world was enchanted and man saw himself as an integral part of it. The complete reversal of this perception in a mere four hundred years or so has destroyed the continuity of the human experience and the integrity of the human psyche. It has very nearly wrecked the planet as well. The only hope, or so it seems to me, lies in a reenchantment of the world.

Here, then, is the crux of the modern dilemma. We cannot go back to alchemy or animism—at least that does not seem likely; but the alternative is the grim, scientistic, totally controlled world of nuclear reactors, microprocessors, and genetic engineering—a world that is virtually upon us already. *Some* type of holistic, or participating, consciousness and a corresponding sociopolitical formation have to emerge if we are to survive as a species. At this point, as I have said, it is not at all evident what this change will involve; but the implication is that a way of life is slowly coming into being which will be vastly different from the epoch that has so deeply colored, in fact created, the details of our lives. Robert Heilbroner has suggested that a time might come, perhaps two hundred years hence, when people will visit the Houston computer center or Wall Street as curious relics of a vanished civilization, but this will necessarily involve a dramatically altered perception of reality.[13] Just as we recognize in a medieval tapestry or alchemical text a world vastly different from our own, so may those people who visit Houston or the tip of Manhattan two centuries

23

from now find our own mental outlook, from the assumptions of nineteenth-century physics to the practice of behavior modification, quite baroque, if not downright incomprehensible.

Willis Harman has called our outlook the "industrial-era paradigm,"[14] but the Industrial Revolution did not begin its "take-off" until the second half of the eighteenth century, whereas the modern paradigm is ultimately the child of the Scientific Revolution. For lack of a better term, then, I shall refer to our world view as the "Cartesian paradigm," after the great methodological spokesman of modern science, René Descartes. I do not wish to suggest that Descartes is the lone architect of our current outlook, but only that modern definitions of reality can be identified with specific planks in his scientific program. To understand the nature and origins of the Cartesian paradigm, then, will be our first task. We shall then be in a position to analyze more closely the nature of the enchanted world view, the historical forces that led to its collapse, and finally the possibilities that exist for a modern and credible form of reenchantment, a cosmos once more our own.

1

The Birth of Modern Scientific Consciousness

[My discoveries] have satisfied me that it is possible to reach knowledge that will be of much utility in this life; and that instead of the speculative philosophy now taught in the schools we can find a practical one, by which, knowing the nature and behavior of fire, water, air, stars, the heavens, and all the other bodies which surround us, as well as we now understand the different skills of our workers, we can employ these entities for all the purposes for which they are suited, and so make ourselves masters and possessors of nature.

—René Descartes, *Discourse on Method* (1637)

Two archetypes pervade Western thinking on the subject of how reality is best apprehended, archetypes that have their ultimate origin in Plato and Aristotle. For Plato sense data were at best a distraction from knowledge, which was the province of un-aided reason. For Aristotle, knowledge consisted in generaliza-tions, but these were derived in the first instance from information gathered from the outside world. These two models of human thinking, termed rationalism and empiricism respectively, formed the major intellectual legacy of the West down to Descartes and Bacon, who represented, in the seventeenth century, the twin poles of epistemology. Yet just as Descartes and Bacon have more in common than apart, so too do Plato and Aristotle. Plato's qual-itative organic cosmos, described in the *Timaeus*, is Aristotle's world as well; and both were seeking the underlying "forms" of the phenomena observed, which were always expressed in teleological terms. Aristotle would not agree with Plato that the "form" of a thing existed in some innate heaven, but nevertheless the reality of, let us say, a discus used at the Olympic games was its Circularity, its Heaviness (inherent tendency to fall to the center of the earth), and so on. This metaphysic was preserved through the Middle Ages, an age noted (from our point of view) for its extensive symbolism. Things were never "just what they were," but always embodied a nonmaterial principle that was seen as the essence of their reality.

Despite the diametrically opposed points of view represented by Bacon's *New Organon* and Descartes' *Discourse on Method*, they possess a commonality that marks them off quite sharply from both the world of the Greeks and that of the Middle Ages. The fundamental discovery of the Scientific Revolution—a discovery epitomized by the work of Newton and Galileo—was that there was no real clash between rationalism and empiricism. The former says that the laws of thought conform to the laws of things; the latter says, always check your thoughts against the data so that you know what thoughts to think. This dynamic relationship between rationalism and empiricism lay at the heart of the Scientific Revolution, and was made possible by the translation of each approach into a concrete tool. Descartes showed that mathematics was the epitome of pure reason, the most trustworthy knowledge available. Bacon pointed out that one had to question nature directly by putting it in a position in which it was forced to yield up its answers. *Natura vexata*, he called it, "nature annoyed": arrange a situation where yes or no must be given in response. Galileo's work illustrates the union of these two tools. For example, roll a ball down an inclined plane and measure distance versus time. Then you will know, precisely, how falling objects behave.

Note that I said *how* they behave, not why. The marriage of reason and empiricism, of mathematics and experiment, expressed this significant shift in perspective. So long as men were content to ask why objects fell, why phenomena occurred, the question of how they fell or occurred was irrelevant. These two questions are not mutually exclusive, at least not in theory; but in historical terms they have proven to be so. "How" became increasingly important, "why" increasingly irrelevant. In the twentieth century, as we shall see, "how" has become our "why."

Viewed from this vantage point, both the *New Organon* and the *Discourse* make for fascinating reading, for we recognize that each

28

author is grappling with an epistemology that has become part of the air we now breathe. Bacon and Descartes interlock in other ways as well. Bacon is convinced that knowledge is power and truth utility; Descartes sees certainty as equivalent to measurement, and wants science to become a "universal mathematics." Bacon's goal, of course, was realized by Descartes' means: precise measurement not only validates or falsifies hypotheses, it also enables the construction of bridges and roads. Hence another crucial seventeenth-century departure from the Greeks: the conviction that the world lies before us to be acted upon, not merely contemplated. Greek thought is static, modern science dynamic. Modern man is Faustian man, an appellation that goes back, even before Goethe, to Christopher Marlowe. Dr. Faustus, sitting in his study ca. 1590, is bored with the works of Aristotle which are spread out before him. "Is to dispute well logic's chiefest end?" he asks himself aloud. "Affords this art no greater miracle? / Then read no more. . . ." [1] In the sixteenth century Europe discovered, or rather decided, that to do is the issue, not to be.

One thing that is conspicuous about the literature of the Scientific Revolution is that its ideologues were self-conscious about their role. Both Bacon and Descartes were aware of the methodological changes taking place, and of the direction in which things would inevitably move. They saw themselves as leading the way, even possibly tipping the balance. Both made it clear that Aristotelianism had had its day. The very title of Bacon's work, *New Organon*, the new instrument, was an attack on Aristotle, whose logic had been, in the Middle Ages, collected under the title *Organon*. Aristotelian logic, specifically the syllogism, had been the basic instrument for apprehending reality, and it was this situation that prompted the complaint of Bacon and Dr. Faustus. Bacon writes that this logic is "no match for the subtlety of nature"; "it gains assent to the proposition, but does not take hold of the thing." Thus it "is idle," he exclaims, "to expect any great advancement in science from the superinducing and engrafting of new things upon old. We must begin anew from the very foundations, unless we would revolve forever in a circle with mean and contemptible progress." [2] Escaping from this circularity involved, as far as Bacon was concerned, a violent shift in perspective, which would lead from the unchecked use of words and reason to the hard data accumulated through the experimental testing of nature. Yet Bacon himself never performed a single experiment, and the method he proposed for ascertaining the truth—compiling tables of data and making generalizations from them—was certainly poorly defined. As a result, historians have erroneously concluded that science grew up "around" Bacon, not through him.[3] Despite the popular conception of the scientific method, most scientists know that truly creative research often begins with wild speculation and flights of fancy that are then subjected to the twin tests of measurement and experiment. Pure Baconianism— expecting results to fall out of the data as if by sheer weight—never really works in practice. Yet this heavily empirical image of Bacon is in fact a result of the nineteenth-century assault on speculation, and the accompanying overemphasis on Bacon's data-collecting side. In the seventeenth and eighteenth centuries, Baconianism was synonymous with the identification of truth with utility, specifically industrial utility. Breaking the Aristotelian-Scholastic circle meant, for Bacon, stepping into the world of the mechanical arts, a step that was literally incomprehensible prior to the mid-sixteenth century. Bacon leaves no doubt that he regards technology as the source of a new epistemology.[4] He tells us that scholarship, which is to say Scholasticism, has stood still for centuries, while technology has made progress; surely it has something to teach us.

> The sciences [he writes] stand where they did and remain almost in the same condition; receiving no noticeable increase.... Whereas in the mechanical arts, which are founded on nature and the light of experience, we see the contrary happen, for these... are continually thriving and growing, as having in them a breath of life.[5]

Natural history, presently understood, says Bacon, is merely the compilation of copious data: descriptions of plants, fossils, and the like. Why should we value such a collection?

> A natural history which is composed for its own sake is not like one that is collected to supply the understanding with information for the building up of philosophy. They differ in many ways, but especially in this: that the former contains the variety of natural species only, and not experiments of the mechanical arts. For even as in the business of life a man's disposition and the secret workings of his mind and affections are better discovered when he is in trouble than at other times; so likewise the secrets of nature reveal themselves more readily under the vexations of art [i.e., artisanry, technology] than when they go their own way. Good hopes may therefore be conceived of natural

philosophy, when natural history, which is the basis and foundation of it, has been drawn up on a better plan; but not till then.[6]

This is a truly remarkable passage, for it suggests for the first time that the knowledge of nature comes about under artificial conditions. Vex nature, disturb it, alter it, anything—but do not leave it alone. Then, and only then, will you know it. The elevation of technology to the level of a philosophy had its concrete embodiment in the concept of the experiment, an artificial situation in which nature's secrets are extracted, as it were, under duress.

It is not that technology was something new in the seventeenth century; the control of the environment by mechanical means in the form of windmills or plows is almost as old as *homo sapiens* himself. But the elevation of this control to a philosophical level was an unprecedented step in the history of human thought. Despite the extreme sophistication of, for example, Chinese technology down to the fifteenth century A.D., it never had occurred to the Chinese (or to Westerners, for that matter) to equate mining or gunpowder manufacture with pure knowledge, let alone with the key to acquiring such knowledge.[7] Science did not, then, grow up "around" Bacon, and his own lack of experimentation is irrelevant. The details of what constituted an experiment were worked out later, in the course of the seventeenth century. The overall framework of scientific experimentation, the technological notion of the questioning of nature under duress, is the major Baconian legacy.

Although it may be reading too much into Bacon, there is a dark hint that the mind of the experimenter, when it adopts this new perspective, will also be under duress. Just as nature must not be allowed to go its own way, says Bacon in the Preface to the work, so it is necessary that "the mind itself be from the very outset not left to take its own course, but guided at every step; and the business be done as if by machinery." To know nature, treat it mechanically; but then your mind must behave mechanically as well.

René Descartes also took his stand against Scholasticism and philosophical verbiage, and felt that nothing less than certainty would do for a true philosophy of nature. The *Discourse*, written some seventeen years after the *New Organon*, is in part an intellectual autobiography. Its author emphasizes the worthlessness of the ancient learning to himself personally, and in doing so implicates the rest of Europe as well. I had the best education France had to offer, he says (he studied at a Jesuit seminary, the Ecole de La Flèche); yet I learned nothing I could call certain. "As far as the opinions which I had been receiving since my birth were concerned, I could not do better than to reject them completely for once in my life time. . . ." [8] As with Bacon, Descartes' goal is not to "engraft" or "superinduce," but to start anew. But how vastly different is Descartes' starting point! It is no use collecting data or examining nature straight off, says Descartes; there will be time enough for that once we learn how to think correctly. Without having a method of clear thinking which we can apply, mechanically and rigorously, to every phenomenon we wish to study, our examination of nature will of necessity be faulty. Let us, then, block out the external world and sort out the nature of right thinking itself.

To start with, says Descartes, it was necessary to disbelieve everything I thought I knew up to this point. This act was not undertaken for its own sake, or to serve some abstract principle of rebellion, but proceeded from the realization that all the sciences were at present on shaky ground. "All the basic principles of the sciences were taken from philosophy," he writes, "which itself had no certain ones." Since my goal was certainty, "I resolved to consider almost as false any opinion which was merely plausible." Thus the starting point of the scientific method, insofar as Descartes was concerned, was a healthy skepticism. Certainly the mind ought to be able to know the world, but first it must rid itself of credulity and medieval rubbish, with which it had become inordinately cluttered. "My whole purpose," he points out, "was to achieve greater certainty and to reject the loose earth and sand in favor of rock and clay."

The principle of methodical doubt, however, brought Descartes to a very depressing conclusion: there was nothing at all of which one could be certain. For all I know, he writes in the *Meditations on First Philosophy* (1641), there could be a total disparity between reason and reality. Even if I assert that God is good, and is not deceiving me when I try to equate reason with reality, how do I know there is not a malignant demon running about who confuses me? How do I know that 2 + 2 do not make 5, and that this demon does not deceive me, every time I make the addition, into believing the numbers add up to 4? But even if this were the case, concludes Descartes, there is one thing I do know: that I exist. For even if I am deceived, there is obviously a "me" who is being deceived. And thus, the bedrock certainty that underlies everything: I think, therefore I am. For Descartes, thinking was identical to existing.

This postulate is, of course, only a beginning. I want to be certain of more than just my own existence. Confronted with the rest of knowledge, however, Descartes finds it necessary to demonstrate (which he does most unconvincingly) the existence of a benevolent Deity. The existence of such a God immediately guarantees the propositions of mathematics, which alone among the sciences relies on pure mental activity. There can be no deception when I sum the angles of a triangle; the goodness of God guarantees that my purely mental operations will be correct, or as Descartes says, clear and distinct. And extrapolating from this, we see that knowledge of the external world will also have certitude if its ideas are clear and distinct, that is, if it takes geometry as its model (Descartes never really did define, to anybody's satisfaction, the terms "clear" and "distinct"). Science, says Descartes, must become a "universal mathematics"; numbers are the only test of certainty.

The disparity between Descartes and Bacon would seem to be complete. Whereas the latter sees the foundations of knowledge in sense data, experiment, and the mechanical arts, Descartes sees only confusion in such subjects and finds clarity in the operations of the mind alone.[9] Thus the method he sets forth for acquiring knowledge is based, he tells us, on geometry. The first step is the statement of the problem that, in its complexity, will be obscure and confused. The second step is breaking the problem down into its simplest units, its component parts. Since one can perceive directly and immediately what is clear and distinct in these simplest units, one can finally reassemble the whole structure in a logical fashion. Now the problem, complex though it may be, is no longer unknown (obscure and confused), because we ourselves have first broken it down and then put it back together again. Descartes was so impressed with this discovery that he regarded it as the key, indeed the only key, to the knowledge of the world. "Those long chains of reasoning," he writes, "so simple and easy, which enabled the geometricians to reach the most difficult demonstrations, had made me wonder whether all things knowable to men might not fall into a similar logical sequence."[10]

Although Bacon's identification of knowledge with industrial utility and his grappling with the concept of experiment based on technology certainly underlie much of our current scientific thought, the implications drawn from the Cartesian corpus exercised a staggering impact on the subsequent history of Western consciousness and (despite the differences with Bacon) served to confirm the technological paradigm—indeed, even helped to launch it on its way. Man's activity as a thinking being—and that is his essence, according to Descartes—is purely mechanical. The mind is in possession of a certain method. It confronts the world as a separate object. It applies this method to the object, again and again and again, and eventually it will know all there is to know. The method, furthermore, is also mechanical. The problem is broken down into its components, and the simple act of cognition (the direct perception) has the same relationship to the knowledge of the whole problem that, let us say, an inch has to a foot: one measures (perceives) a number of times, and then sums the results. Subdivide, measure, combine; subdivide, measure, combine.

This method may properly be called "atomistic," in the sense that knowing consists of subdividing a thing into its smallest components. The essence of atomism, whether material or philosophical, is that a thing consists of the sum of its parts, no more and no less. And Descartes' greatest legacy was surely the mechanical philosophy, which followed directly from this method. In the *Principles of Philosophy* (1644) he showed that the logical linking of clear and distinct ideas led to the notion that the universe was a vast machine, wound up by God to tick forever, and consisting of two basic entities: matter and motion. Spirit, in the form of God, hovers on the outside of this billiard-ball universe, but plays no direct part in it. All nonmaterial phenomena ultimately have a material basis. The action of magnets, attracting each other over a distance, may seem to be nonmaterial, says Descartes, but the application of the method can and will ultimately uncover a particulate basis for their behavior. What Descartes does, really, is provide Bacon's technological paradigm with strong philosophical teeth. The mechanical philosophy, the use of mathematics, and the formal application of his four-step method enable the manipulation of the environment to take place with some sort of logical regularity.

The identification of human existence with pure ratiocination, the idea that man can know all there is to know by way of his reason, included for Descartes the assumption that mind and body, subject and object, were radically disparate entities. Thinking, it would seem, separates me from the world I confront. I perceive my body and its functions, but "I" am not my body. I can learn about the (mechanical) behavior of my body by applying the

Cartesian method—and Descartes does this in his treatise *On Man* (1662)—but it always remains the object of my perception. Thus Descartes depicted the operation of the human body by means of analogy to a water fountain, with mechanical reflex action being the model of most, if not all, human behavior. The mind, *res cogitans* ("thinking substance"), is in a totally different category from the body, *res extensa* ("extended substance"), but they do have a mechanical interaction that we can diagram as in Figure 3, below. If the hand touches a flame, the fire particles attack the finger, pulling a thread in the tubular nerve which releases the "animal spirits" (conceived as mechanical corpuscles) in the brain. These then run down the tube and jerk the muscles in the hand.[11]

There is, it seems to me, an uncanny similarity between this diagram and that of Laing's "false-self system" depicted in the Introduction (see Figure 2). Schizophrenics typically regard their bodies as "other," "not-me." In Descartes' diagram, too, brain (inner self) is the detached observer of the parts of the body; the interaction is mechanical, as though one saw oneself behaving as a robot—a perception that is easily extended to the rest of the world. To Descartes, this mind-body split was true of *all* perception and behavior, such that in the act of thinking one perceived oneself as a separate entity "in here" confronting things "out there." This schizoid duality lies at the heart of the Cartesian paradigm.

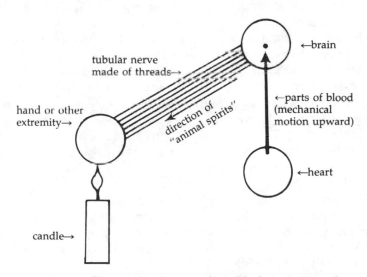

Figure 3. Descartes' conception of mind-body interaction.

Descartes' emphasis on clear and distinct ideas, and his basing of knowledge on geometry, also served to reaffirm, if not actually canonize, the Aristotelian principle of non-contradiction. According to this principle, a thing cannot both be and not be at the same time. When I strike the letter "A" on my typewriter, I get an "A" on the paper (assuming the machine is working properly), *not* a "B." The cup of coffee sitting to the right of me could be put on a scale and found to have a weight of, say, 5.24 ounces, and this fact means that the object does *not* weigh ten pounds or two grams. Since the Cartesian paradigm recognizes no self-contradictions in logic, and since logic (or geometry), according to Descartes, is the way nature behaves and is known to us, the paradigm allows for no self-contradictions in nature.

The problems with Descartes' view are perhaps obvious, but for now, it will suffice to note that real life operates dialectically, not critically.[12] We love and hate the same thing simultaneously, we fear what we most need, we recognize ambivalence as a norm rather than an aberration. Descartes' devotion to critical reason led him to identify dreams, which are profoundly dialectical statements, as the model of unreliable knowledge. Dreams, he tells us in the *Meditations on First Philosophy,* are not clear and distinct, but invariably obscure and confused. They are filled with frequent self-contradictions, and possess (from the viewpoint of critical reason) neither internal nor external coherence. For example, I might dream that a certain person I know is my father, or even that I am my father, and that I am arguing with him. But this dream is (from a Cartesian point of view) internally incoherent, because I am simply not my father, nor can he be himself and someone else as well; and it is externally incoherent, because upon waking, no matter how real it all seems for a moment, I soon realize that my father is three thousand miles away, and that the supposed confrontation never took place. For Descartes, dreams are not material in nature, cannot be measured, and are not clear and distinct. Given Descartes' criteria, then, they contain no reliable information.

In summation, rationalism and empiricism, the twin poles of knowledge so strongly represented in Descartes and Bacon respectively, can be regarded as complementary rather than irrevocably conflicting. Descartes, for example, was hardly opposed to experiment when it served to adjudicate between rival hypotheses—a role it retains to this day. And as I have argued, his atomistic approach, and his emphasis on material reality and its measurement easily lent themselves to the sort of knowledge and economic power that Bacon envisaged as possible for England and Western Europe. Still, this synthesis of reason and empiricism lacked a concrete embodiment, a clear demonstration of how the new methodology might work in practice; the scientific work of Galileo and Newton provided precisely such a demonstration. These men were concerned not merely with the question of methodological exposition (though each certainly made his own contribution to that subject), but sought to illustrate exactly how the new methodology could analyze the simplest events: the stone falling to earth, the ray of light passing through a prism. Through such specific examples the dreams of Bacon and Descartes were translated into a working reality.

Galileo, in his painstaking studies of motion carried out in the twenty years preceding the publication of the *New Organon*, had already made explicit what Bacon only implied as an artificial construct in his generalizations about the experimental method.[13] Frictionless planes, massless pulleys, free-fall with zero air resistance—all of these "ideal types" that form the basic problem sets in freshman physics are the legacy of that Italian genius, Galileo Galilei. Galileo is popularly remembered for an experiment he never performed—dropping weights from the Leaning Tower of Pisa—but in fact he conducted a far more ingenious experiment on falling objects—an experiment that exemplifies many of the major themes of modern scientific inquiry. The belief that large or dense objects should strike the ground faster than light ones follows as a direct consequence of Aristotle's teleological physics, and was widely held throughout the Middle Ages. If things fall to the ground because they seek their "natural place," the earth's center, we can see why they would accelerate as they approach it. They are excited, they are coming home, and like all of us they speed up as they approach the last leg of the journey. Heavy objects drop a given distance in a shorter time than light ones because there is more matter to become excited, and thus they attain a higher speed and strike the ground first. Galileo's argument, that a very large object and a very small one would make the drop in the same time interval, was based on an assumption that could neither be proven nor falsified: that falling objects are inanimate and thus have neither goals nor purposes. In Galileo's scheme of things, there is no "natural place" anywhere in the universe. There is but matter and motion, and we can but observe and measure it. The proper subject for the investigation of nature, in other words, is not why an object falls—there is no *why*—but *how*; in this case, how much distance in how much time.

Although Galileo's assumptions may seem obvious enough to us, we must remember how radically they violated not only the common-sense assumptions of the sixteenth century, but common-sense observations in general. If I look around, and see that I am rooted to the ground, and that objects released in midair fall to the floor, isn't it perfectly reasonable to regard "down" as their natural, that is to say inherent, motion? In his studies of childhood cognition, Swiss psychologist Jean Piaget discovered that until about age seven at the latest, children are Aristotelians.[14] When asked why objects fell to the floor, Piaget's subjects replied, "because that is where they belong" (or some variation of this idea). Perhaps most adults are emotional Aristotelians as well. Aristotle's proposition that there is no motion without a mover, for example, seems instinctively correct; and most adults, when asked to react immediately to the notion, will affirm it. Galileo refuted the proposition by rolling a ball down two inclined planes, juxtaposed as in Figure 4:

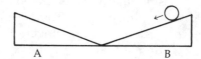

Figure 4. Galileo's experiment for showing that motion does not require a mover.

The ball rolls down B and up A, but not to quite the same height from which it began. Then it rolls back down A and up B, again losing height; back and forth, back and forth, until the ball finally settles in the "valley" and comes to rest. If we polish the planes, making them smoother and smoother, the ball stays in motion for a longer and longer period of time. In the limiting case, where friction = 0, the motion would go on forever: hence, motion without a mover. But there is one problem with Galileo's argument: *there is no limiting case.* Ther are no frictionless planes. The law of inertia may state that a body continues in motion or in a state of rest unless acted upon by an outside force, but in fact, in the case of motion, there is *always* an outside force, if nothing more than the friction between the object and the surface over which it moves.[15]

The experiment Galileo designed to measure distance against time was a masterpiece of scientific abstraction. To drop weights

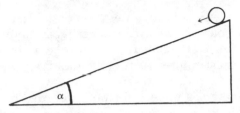

Figure 5. Galileo's experiment for deriving the law of free-fall.

from the Leaning Tower, Galileo realized, was absolutely useless. Simon Stevin, the Dutch physicist, had tried a free-fall experiment in 1586 only to learn that the speed was too fast for measurement. Thus, said Galileo, I shall "dilute" gravity by rolling a ball down an inclined plane, made as smooth as possible to reduce friction. If we were to make the slope steeper by increasing the angle α, as in Figure 5, we would reach the free-fall situation that we seek to explore at the limiting case, in which $\alpha = 90$ degrees. Hence let us take a smaller angle, say $\alpha = 10$ degrees, and let it serve as an approximation. Galileo first used his pulse as a timer, and later a bucket of water with a hole in it which permitted the water to drip at regular intervals. By running a series of trials, he finally came up with a numerical relationship, that distance is proportional to the square of the time. In other words, if an object—any object, light or heavy—falls a unit distance in one second, then it falls a distance of four times that in two seconds, nine times that in three seconds, and so on. In modern terminology, $s = kt^2$, where s is distance, t time, and k a constant.

Both of these inclined plane experiments illustrate the highly ingenious combination of rationalism and empiricism which was Galileo's trademark. Consult the data, but do not allow them to confuse you. Separate yourself from nature so you can, as Descartes would later urge, break it into the simplest parts and extract the essence—matter, motion, measurement. In general terms, Galileo's was not an altogether new contribution to human history, as we shall see in Chapter 3; but it did represent the final stage in the development of nonparticipating consciousness, that state of mind in which one knows phenomena precisely in the act of distancing oneself from them. The notion that nature is alive is clearly a stumbling block to this mode of understanding. For when we regard material objects as extensions of ourselves (alive, endowed with purpose) and allow ourselves to be distracted by the sensuous details of nature, we are powerless to control nature, and thus, from Galileo's point of view, can never really know it. The new science enjoins us to step outside of nature, to reify it, reduce it to measurable Cartesian units; only then can we have definitive knowledge of it. As a result—and Galileo was not interested in ballistics and materials science for nothing—we shall supposedly be able to manipulate it to our advantage.

Clearly, the identification of truth with utility was closely allied to the Galilean program of nonparticipating consciousness and the shift from "why" to "how." Unlike Bacon, Galileo did not make this identification explicit, but once natural processes are stripped of immanent purpose, there is really nothing left in objects but their value for something, or someone, else. Max Weber called this attitude of mind *zweckrational*, that is, purposively rational, or instrumentally rational. Embedded within the scientific program is the concept of manipulation as the very touchstone of truth. To know something is to control it, a mode of cognition that led Oskar Kokoschka to observe that by the twentieth century, reason had been reduced to mere function.[16] This identification, in effect, renders all things meaningless, except insofar as they are profitable or expedient; and it lies at the heart of the "fact-value distinction," briefly discussed in the Introduction. The medieval Thomistic (Christian-Aristotelian) synthesis, that saw the good and the true as identical, was, in the first few decades of the seventeenth century, irrevocably dismantled.

Of course, Galileo did not regard his method as merely useful, or heuristically valuable, but uniquely true, and it was this epistemological stance that created havoc with the church. For Galileo, science was not a tool, but the one path to truth. He tried to keep its claims separate from those of religion, but failed: the church's historical commitment to Aristotelianism proved to be too great. In this conflict Galileo, as a good Catholic, was understandably worried that the church, by insisting on its infallibility, would inevitably deal itself a serious blow. Galileo's life, in fact, is the story of the prolonged struggle, and failure, to win the church over to the cause of science; and in his play *Galileo*, Bertolt Brecht makes this theme of the irresistibility of the scientific method central to the story. He has Galileo wander through the drama carrying a pebble, which he occasionally drops to illustrate the power of sensory evidence. "If anybody were to drop a stone," he asks his friend Sagredo, "and tell [people] that it didn't fall, do you think they would keep quiet? The evidence of your own eyes is a very seductive thing. Sooner or later everybody must succumb to it." And Sagredo's reply? "Galileo, I am helpless when you talk."[17] The logic of science had a historical logic as well. In time all alternative methodologies animism, Aristotelianism, or argument by papal fiat—crumbled before the seductiveness of free rational inquiry.

The lives of Newton and Galileo stretch across the whole of the seventeenth century, for the former was born in the same year that the latter died, 1642, and together they embrace a revolution in human consciousness. By the time of Newton's death in 1727, the educated European had a conception of the cosmos, and of the nature of "right thinking," which was entirely different from that of his counterpart of a century before. He now regarded the earth as revolving around the sun, not the reverse;[18] believed that all phenomena were constituted of atoms, or corpuscles, in motion and susceptible to mathematical description; and saw the solar system as a vast machine, held together by the forces of gravity. He had a precise notion of experiment (or at least paid lip service to it), and a new notion of what constituted acceptable evidence and proper explanation. He lived in a predictable, comprehensible, yet (in his own mind) very exciting sort of world. For in terms of material control, the world was beginning to exhibit an infinite horizon and endless opportunities.

More than any other individual, Sir Isaac Newton is associated with the scientific world view of modern Europe. Like Galileo, Newton combined rationalism and empiricism into a new method; but unlike Galileo, he was hailed by Europe as a hero rather than having to recant his views and spend his mature years under house arrest. Most important, the methodological combination of reason and empiricism became, in Newton's hands, a whole philosophy of nature which he (unlike Galileo) was successful in stamping upon Western consciousness at large. What made the eighteenth century *the* Newtonian century was the solution to the problem of planetary motion, a problem that, it was commonly believed, not even the Greeks had been able to solve (the Greeks, it should be noted, took a more positive view of their own achievement). Bacon had derided the ancient learning, but he did not speak for the majority of Europeans. The strong revival of classical learning in the sixteenth century, for example, reflected the belief that despite the enormous problems with the Greek cosmological model, their epoch was and would remain the true Golden Age of mankind. Newton's precise mathematical description of a heliocentric solar system changed all that; he not only summed up the universe in four simple algebraic formulas, but he also accounted for hitherto unexplained phenomena, made accurate predictions, clarified the relation between theory and experiment, and even sorted out the role of God in the whole system. Above all, Newton's system was atomistic: the earth and sun, being composed of atoms themselves, behaved in the same way that any two atoms did, and vice versa. Thus both the smallest and the largest objects in the universe were seen to obey identical laws. The moon's relationship to the earth was the same as that of a falling apple. The mystery of nearly two millennia was over: one could be reassured that the heavens that confront us on a starry night held no more secrets than a few grains of sand running through our fingers.

Newton deliberately titled his major work, popularly called the *Principia*, the *Mathematical Principles of Natural Philosophy* (1686),[19] the two adjectives serving to emphasize his rejection of Descartes, whose *Principles of Philosophy* he regarded as a collection of unproven hypotheses. Step by step he analyzed Descartes' propositions about the natural world and demonstrated their falsity. For example, Descartes envisaged the matter of the universe circulating in whirlpools, or vortices. Newton was able to show that this theory contradicted the work of Kepler, which seemed quite reliable; and that if one experimented with models of vortices by spinning buckets of fluid (water, oil, pitch), the contents would eventually slow down and stop, indicating that on Descartes' hypothesis the universe would have come to a standstill long ago. Despite his attacks on Descartes' views, it is clear from recent research that Newton was a Cartesian right up to the publication of the *Principia*; and when one reads the work, one is struck by an awesome fact: Newton made the Cartesian world view tenable by falsifying all of its details. In other words, although Descartes' facts were wrong and his theories insupportable, the central Cartesian outlook—that the world is a vast machine of matter and motion obeying mathematical laws—was thoroughly validated by Newton's work. For all of Newton's brilliance, the real hero (some would say ghost) of the Scientific Revolution was René Descartes.

But Newton did not have his triumph so easily. His entire view of the cosmos hinged on the law of universal gravitation, or gravity, and even after it had been given an exact mathematical formulation, no one knew just what this attraction was. Cartesian thinkers pointed out that their own mentor had wisely restricted himself to motion by direct impact, and ruled out what scientists would later call action-at-a-distance. Newton, they argued, has not *explained* gravity, but merely stated its effects, and thus it really is, in his system, an occult property. Where is this "gravity" that he makes so much of? It can be neither seen, nor heard, nor felt, nor smelled. It is, in short, as much a fiction as the vortices of Descartes.

Privately, Newton agonized over this judgment. He felt that his critics were correct. Early in 1692 or 1693 he wrote his friend the Reverend Richard Bentley the following admission:

> That gravity should be innate, inherent and essential to matter, so that one body may act upon another at a distance through a *vacuum*, without the mediation of anything else, by and through which their action and force may be conveyed from one to another, is to me so great an absurdity that I believe no man who has in philosophical matters a competent faculty of thinking can ever fall into it. Gravity must be caused by an agent acting constantly according to certain laws, but whether this agent be material or immaterial I have left to the consideration of my readers.[20]

Publicly, however, Newton adopted a stance that established, once and for all, the philosophical relationship between appearance and reality, hypothesis and experiment. In a section of the *Principia* entitled "God and Natural Philosophy," he wrote:

> Hitherto we have explained the phenomena of the heavens and of our sea by the power of gravity, but have not yet assigned the cause of this power. This is certain, that it must proceed from a cause that penetrates to the very centers of the sun and planets. . . . But hitherto I have not been able to discover the cause of those properties of gravity from phenomena, and I frame no hypotheses; for whatever is not deduced from the phenomena is to be called a hypothesis, and hypotheses, whether metaphysical or physical, whether of occult qualities or mechanical, have no place in experimental philosophy.[21]

Newton was echoing the major theme of the Scientific Revolution: our goal is how, not why. That I cannot explain gravity is irrelevant. I can measure it, observe it, make predictions based on it, and this is all the scientist has to do. If a phenomenon is not measurable, it can "have no place in experimental philosophy." This philosophical position, in its various forms called "positivism," has been the public face of modern science down to the present day.[22]

The second major aspect of Newton's work was best delineated

43

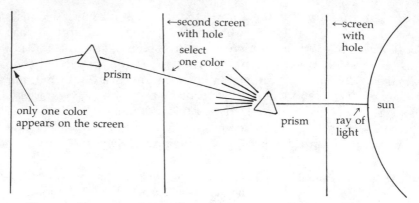

Figure 6. Newton's subdivision of white light into monochromatic rays.

in the *Opticks* (1704), in which he was able to wed philosophical atomism to the definition of experiment which had been crystallizing in the minds of scientists throughout the previous century. As a result, Newton's researches on light and color became the model for the correct analysis of natural phenomena. The question was, is white light simple or complex? Descartes, for one, had regarded it as simple, and saw colors as the result of some sort of modification of the light. Newton believed white light was in fact composed of colors that somehow cancelled each other out in combination to produce the effect of white. How to decide between these two claims?

In the experiment illustrated in Figure 6, Newton took white light, broke it into parts with a prism, selected one of the parts, and showed that it could not be further broken down. He did this with each color, demonstrating that monochromatic light could not be subdivided. Next, Newton ran the experiment in the opposite direction: he broke the ray of white light into its parts, and then recombined them by passing them through a convex lens (see Figure 7). The result was white light. This atomistic approach, which follows Descartes' four-step method exactly, establishes the thesis beyond doubt. But as in the case of gravity, the Cartesians took issue with Newton. Where, they asked, is your *theory* of light and color, where is your *explanation* of this behavior? And as in the previous case, Newton retreated behind the smokescreen of positivism. I am looking for laws, or optical facts, he replied, not hypotheses. If you ask me what "red" is, I can only tell you that it

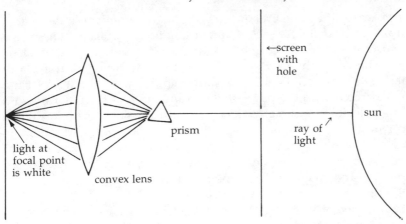

Figure 7. Newton's recombination of monochromatic light rays into white light.

is a number, a certain degree of refrangibility, and the same is true for each of the other colors. I have measured it: that is enough.

In this case too, of course, Newton struggled with possible explanations for the behavior of light, but the combination of (philosophical) atomism, positivism, and experimental method—in short, the definition of reality—is still very much with us today. To know something is to subdivide it, quantify it, and recombine it; is to ask "how," and never get entangled in the complicated underbrush of "why." It is, above all, to distance yourself from it, as Galileo pointed out; to make it an abstraction. The poet may get uncritically effusive about a red streak across the sky as the sun is going down, but the scientist is not so easily deluded: he knows that his emotions can teach him nothing substantial. The red streak is a number, and that is the essence of the matter.

To summarize our discussion of the Scientific Revolution, it is necessary to note that in the course of the seventeenth century Western Europe hammered out a new way of perceiving reality. The most important change was the shift from quality to quantity, from "why" to "how." The universe, once seen as alive, possessing its own goals and purposes, is now a collection of inert matter, hurrying around endlessly and meaninglessly, as Alfred North Whitehead put it.[23] What constitutes an acceptable explanation has thus been radically altered. The acid test of existence is quantifiability, and there are no more basic realities in any object than the parts

45

into which it can be broken down. Finally, atomism, quantifiability, and the deliberate act of viewing nature as an abstraction from which one can distance oneself—all open the possibility that Bacon proclaimed as the true goal of science: control. The Cartesian or technological paradigm is, as stated above, the equation of truth with utility, with the purposive manipulation of the environment. The holistic view of man as a part of nature, as being at home in the cosmos, is so much romantic claptrap. Not holism, but domination of nature; not the ageless rhythm of ecology, but the conscious management of the world; not (to take the process to its logical end point) "the magic of personality, [but] the fetishism of commodities."[24] In the mind of the eighteenth and nineteenth centuries, medieval man (or woman) had been a passive spectator of the physical world. The new mental tools of the seventeenth century made it possible to change all that. It was now within our power to have heaven on earth; and the fact that it was a material heaven hardly made it less valuable.

Nevertheless, it was the Industrial Revolution that put the Scientific Revolution on the map. Bacon's dream of a technological society was not realized in the seventeenth century or even in the eighteenth, although things were beginning to change by 1760. Ideas, as we have said, do not exist in a vacuum. People could regard the mechanical world view as the true philosophy without feeling compelled to transform the world according to its dictates. The relationship between science and technology is very complicated, and it is in fact in the twentieth century that the full impact of the Cartesian paradigm has been most keenly felt. To grasp the meaning of the Scientific Revolution in Western history we must consider the social and economic milieu that served to sustain this new way of thinking. The sociologist Peter Berger was correct when he said that ideas "do not succeed in history by virtue of their truth but by virtue of their relationships to specific social processes."[25] Scientific ideas are no exception.

2

Consciousness and Society in Early Modern Europe

From whence there may arise many admirable advantages, towards the increase of the *Operative,* and the *Mechanick* Knowledge, to which this Age seems so much inclined, because we may perhaps be inabled to discern all the secret workings of Nature, almost in the same manner as we do those that are the productions of Art, and are manag'd by Wheels, and Engines, and Springs, that were devised by humane Wit.

—Robert Hooke, *Micrographia* (1665)

The collapse of a feudal economy, the emergence of capitalism on a broad scale, and the profound alteration in social relations that accompanied these changes provided the context of the Scientific Revolution in Western Europe. The equating of truth with utility, or cognition with technology, was an important part of this general process. Experiment, quantification, prediction and control formed the parameters of a world view that made no sense within the framework of the medieval social and economic order. The individuals discussed in Chapter 1 would not have been possible in an earlier age; or, perhaps more to the point, would have been ignored, as were Roger Bacon and Robert Grosseteste, who pioneered the experimental method in the thirteenth century. Modern science, in short, is the mental framework of a world defined by capital accumulation, and ultimately, to quote Ernest Gellner, it became the "mode of cognition" of industrial society.[1]

It is not my intention to argue that capitalism "caused" modern science. The relationship between consciousness and society has always been problematic because all social activities are permeated by ideas and attitudes and there is no way to analyze society in a strictly functional way.[2] We are confronted, then, with a structural totality, or historical gestalt, and my point in this chapter will be that science and capitalism form such a unit. Science acquired its factual and explanatory power only within a context that was "congruent" to those facts and explanations. It will be necessary, therefore, to look at science as a system of thought adequate to a certain historical epoch; to try to separate ourselves from the common impression that it is an absolute, transcultural truth.[3]

Let us begin our examination of this theme by comparing the Aristotelian and seventeenth-century world views, and then consider the changes wrought by the Commercial Revolution of the fifteenth and sixteenth centuries on the social and economic world of feudalism (see Chart 1).

The most striking aspect of the medieval world view is its sense of closure, its completeness. Man is at the center of a universe that is bounded at its outermost sphere by God, the Unmoved Mover. God is the one entity that, in Aristotle's terminology, is pure actuality. All other entities are endowed with purpose, being partly actual and partly potential. Thus it is the goal of fire to move up, of

Chart 1. Comparison of world views

World view of the Middle Ages	*World view of the seventeenth century*
Universe: geocentric, earth in the center of a series of concentric, crystalline spheres. Universe closed, with God, the Unmoved Mover, as the outermost sphere.	*Universe:* heliocentric; earth has no special status, planets held in orbit by gravity of the sun. Universe infinite.
Explanation: in terms of formal and final causes, teleological. Everything but God in process of Becoming; natural place, natural motion.	*Explanation:* strictly in terms of matter and motion, which have no higher purposes. Atomistic in both the material and philosophical sense.
Motion: forced or natural, requires a mover.	*Motion:* to be described, not explained; law of inertia.
Matter: continuous, no vacua.	*Matter:* atomic, implying existence of vacua.
Time: cyclical, static.	*Time:* linear, progressive.
Nature: understood via the concrete and the qualitative. Nature is alive, organic; we observe it and make deductions from general principles.	*Nature:* understood via the abstract and quantitative. Nature is dead, mechanistic, and is known via manipulation (experiment) and mathematical abstraction.

earth (matter) to move down, and of species to reproduce themselves. Everything moves and exists in accordance with divine purpose. All of nature, rocks as well as trees, is organic and repeats itself in eternal cycles of generation and corruption. As a result, this world is ultimately changeless, but being riddled with purpose, is an exceptionally meaningful one. Fact and value, epistemology and ethics, are identical. "What do I know?" and "How should I live?" are in fact the same question.

Turning to the world view of the seventeenth century, we are apt to note first of all the absence of any immanent meaning. As E. A. Burtt describes it, the seventeenth century, which began with the search for God in the universe, ended by squeezing Him out of it altogether.[4] Things do not possess purpose, which is an anthropocentric notion, but only behavior, which can (and must) be described in an atomistic, mechanical, and quantitative way. As a result, our relationship to nature is fundamentally altered. Unlike medieval man, whose relationship with nature was seen as being reciprocal, modern man (existential man) sees himself as having the ability to control and dominate nature, to use it for his own purposes. Medieval man was given a purposeful position in the universe; it did not require an act of will on his part. Modern man, on the other hand, is enjoined to find his own purposes. But what those purposes are or should be cannot, for the first time in history, be logically derived. In short, modern science is grounded in a sharp distinction between fact and value; it can only tell us how to do something, not what to do or whether we should do it.

The openness that we see as characteristic of seventeenth-century consciousness is also antithetical to the medieval cosmos. The universe has become infinite, motion (change) is a given, and time is linear. The notion of progress and the sense that activity is cumulative characterize the world view of early modern Europe.

Finally, what is "really" real for the seventeenth century is what is abstract. Atoms are real, but invisible; gravity is real, but, like momentum and inertial mass, can only be measured. In general, abstract quantification serves as explanation. It was this loss of the tangible and meaningful that drove the more sensitive minds of the age—Blaise Pascal and John Donne, for example—to the edge of despair. The "new Philosophy calls all in doubt," wrote the latter in 1611; "Tis all in peeces, all cohaerance gone." Or in Pascal's phrase, "the silences of the infinite spaces terrify me."[5]

The culture that was permeated by the Aristotelian world view

51

was, as we know, characterized by a feudal economy and a religious way of life. By and large, food and handicrafts were produced not for market and profit, but for immediate consumption and use. Excepting luxury items, trade existed only within local areas, and more closely resembled the tribute structure of the ancient Roman Empire (out of whose disintegration feudalism arose) than our modern notion of commercial exchange. Until the late fifteenth century almost all shipping was coastal: boats stayed within sight of land for fear of getting lost. The guilds, which produced for personal commission, emphasized quality rather than quantity, and closely guarded the secrets of craftsmanship. There was no notion of mass production, and very little division of labor. The economy was, essentially, a self-contained reward system. It could not be described as "going" anywhere, and, in general, our notions of growth and expansion would have made little sense in this static and self-sufficient world. In the Middle Ages, meaning was assured, both politically and religiously. The church was the ultimate reference when one sought to explain a phenomenon, whether it occurred in nature or in human life. Furthermore, the social order made sense in a direct and personal way. Justice and political power were administered in terms of loyalty and attachment—vassal to lord, serf to land, apprentice to master—and the system, as a result, possessed few abstractions. If the Middle Ages seem, from our vantage point, to be hermetically sealed, they had the advantage (despite the extreme instability afforded by plague and natural disaster) of being psychologically reassuring to their inhabitants.[6]

It was, however, in the economic sphere that the feudal system became increasingly nonviable. In terms of economic payoff, the limits of feudalism had been reached as early as the thirteenth century. Since significant capital investment in agriculture was not forthcoming, there existed an upper limit to productivity. This limit in turn caused a strain that was starting to transform peasant rebellions that had begun in the thirteenth century into a class war. In response to this threat, there emerged an enormous pressure to expand the geographical base of economic operations. New areas for the cultivation of sugar and wheat, direct access to the spices that could disguise bad meat, new sources of wood, and more extensive fishing grounds were all seen as necessary to the survival of European civilization. In addition, the fall of Constantinople in 1453 gave the Ottoman Turks hegemony over Eastern trade, creating the need for a non-Mediterranean passage to the East. All these factors contributed to the rapid ascendancy of the imperial program of expansion, and with this interest came a host of inventions that made such a program possible. The full-rigged ship appeared, better able to harness the wind. In the sixteenth century the English set cannon in portholes for easier maneuverability. Gunpowder, which the ancient Chinese had invented and used for fireworks displays, became the basis for the firearm industry. It was no accident that Francis Bacon identified the compass and gunpowder as the twin keys to naval hegemony. The first maps designed with compass knowledge—the beautiful *portolani* still preserved in the libraries of major European cities—began to appear, as did new models of the globe. The image of boats hugging the coast, almost a perfect metaphor for the tight mental horizon of the Middle Ages, was crumbling. It was now the age of Magellan and Columbus and Vasco da Gama. The expansion of consciousness, and territory, made the closed medieval cosmos seem increasingly quaint.

Concomitant to, and directly following on, the Commerical Revolution was a series of developments which smashed the feudal system and established the capitalist mode of production in Western Europe. Commerce naturally began to influence industry. The Commercial Revolution, with its sharply increased volume of long-distance trading, broke down the personal relationship between guild master and customer. If the former were to sell to distant markets, he needed merchant help and credit. The merchant first obtained exclusive disposal of the manufacturer's output, and later began to advance the artisan money on raw materials. Eventually, the artisan fell into such debt that he had to turn his shop over to the merchant, who became a merchant-manufacturer, or entrepreneur. The same process that destroyed guild-master and journeyman turned the peasant into a wage earner. In fifteenth-century England, the rise of the rural "putting-out" system (domestic industry), especially in textile manufacture, marked the beginning of a shift of capital investment away from the cities. Peasants began to devote their energies to various aspects of cloth production, and the cloth guilds began to fail as a result.

The Commercial Revolution also generated profits from trade which could be invested in agriculture and manufacturing. Some industries, such as mining, book printing, shipbuilding (which

53

now employed thousands), and the manufacture of cannon, required great capital outlay from the start, and thus could not be contained within the narrow world of craft production. In some cases, especially when the product had a military use, the state itself became the leading customer. State arsenals, such as the great arsenal at Venice, the scene of much of Galileo's research, became major manufacturing centers in themselves. Military manufacture also had close ties to mining and metallurgy, which expanded dramatically in the early modern period. The application of water power to mining, and the creation of a new type of forge, made possible the casting of guns. A host of technical improvements for pumping, ventilating, and driving mechanisms was developed—and illustrated in lavish detail in such books.as Biringuccio's *Pirotechnia* (1540) and Agricola's *De Re Metallica* (1556). England in particular experienced both industrial growth and commercial expansion after 1550. She began casting cannon in iron (since she lacked bronze); introducing such industries as paper, gunpowder, alum, brass, and saltpeter; substituting coal for wood; introducing new techniques in mining and metallurgy; and squeezing the Hanseatic merchants out of the textile market.

There was no way that the medieval Christian-Aristotelian synthesis could withstand such revolutionary changes, and if we consult the characteristics of the seventeenth-century world view listed earlier in this chapter, we find the counterpart to the economic transformation just described. Heliocentricity reflects not only the awareness that the universe is infinite, but also the European discovery of other worlds and the consequent loss of the sense of European uniqueness. In his *On the Revolution of the Celestial Orbs* (1543), Copernicus cites the widening of geographical horizons as a major influence on his thinking. Turning to the category of explanation, we see that explanations of events are now couched in terms of the mechanical, and mathematically describable, motion of inert matter. Nature (including human beings) is seen as so much stuff to be grasped and shaped. Nothing can have purpose in itself, and values—as Machiavelli was among the first to argue—are just so much sentiment. Reason is now completely (at least in theory) instrumental, *zweckrational*. One can no longer ask, "Is this good?," but only, "Does this work?," a question that reflects the mentality of the Commercial Revolution and the growing emphasis on production, prediction, and control.

Because we ourselves live in a society so completely dominated by a money economy, because the cash value of things has become their only value, it is difficult for us to imagine an age not ruled by money and almost impossible to understand the formative influence that the introduction of a money economy exerted on the consciousness of early modern Europe. The sudden emphasis on money and credit was the most obvious fact of economic life during the Renaissance. The accumulation of vast sums in the hands of single individuals, like the Medici, gave capital a magical quality, the more so as the increasingly popular sale of indulgences brought entry into heaven under its sway. Salvation had literally been the goal of Christian life; now, since it could be purchased, money was. This penetration of finance into the very core of Christianity could not help but rupture the Thomistic synthesis. The German sociologist Georg Simmel argued that the money economy "created the ideal of exact numerical calculation," and that the "mathematically exact interpretation of the cosmos" was the "theoretical counterpart of a money economy." In a society that was coming to regard the world as one big arithmetical problem, the notion that there existed a sacred relationship between the individual and the cosmos seemed increasingly dubious.[7]

Money's seemingly unlimited ability to reproduce itself further substantiated the notion of an infinite universe which was so central to the new world view. Profit, the crux of the capitalist system, is open ended. A "capitalist economy and modern methodical science," wrote the historian Alfred von Martin,

> are the expression of an urge towards what is on principle unlimited and without bounds; they are the expression of a dynamic will to progress *ad infinitum*. Such were the inevitable consequences of the breakup of an economically as well as intellectually closed community. Instead of a closed economy administered in the traditional mode and by a privileged group by way of monopoly, we now find an open cycle and the corresponding change in consciousness.[8]

The emphasis on individual will which we identify with Renaissance thought, specifically with the merchant-entrepreneurial class, also had an obvious affinity with the new arithmetical *Weltanschauung*. The same class that came to power through the new economy, that glorified the effort of the individual, and that began to see in financial calculation a way of comprehending the entire cosmos, came to regard quantification as the key to personal success because quantification alone was thought to enable mastery over nature by a rational understanding of its laws. Both money and scientific intellect (especially in its Cartesian identification with mathematics) have a purely formal, and thus "neutral" aspect. They have no tangible content, but can be bent to any purpose. Ultimately, they became the purpose. Historically, the circle was thus complete, as Figure 8 illustrates:

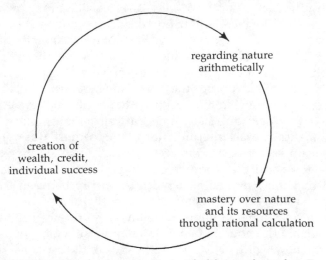

Figure 8. The new cycle of economic/scientific life in early modern Europe.

Finally, even the notion of time—and few things are as basic to human consciousness as the way in which the passage of events is perceived—underwent a fundamental transformation. As Mircea Eliade points out in *The Myth of the Eternal Return*, the premodern conception of time is cyclical. For the people of the Middle Ages, the seasons and events of life followed one another with a comforting regularity. The notion of time as linear was experientially alien to this world, and the need to measure it correspondingly muted. But by the thirteenth century this situation was already changing. Time, wrote Alfred von Martin,

> was felt to be slipping away continuously.... After the thirteenth century the clocks in the Italian cities struck all the twenty-four hours of the day. It was realized that time was always short and hence valuable, that one had to husband it and use it economically if one wanted to become the "master of all things." Such an attitude had been unknown to the Middle Ages; to them time was plentiful and there was no need to look upon it as something precious.[9]

56

The new concern with time running out was much in evidence by the sixteenth century. The phrase "time is money" dates from this period, as does the invention of the pocket watch, in which time, like money, could be held in the hand or pocket. The mentality that seeks to grasp and control time was the same mentality that produced the world view of modern science. Western industrial nations have pushed this change in attitude to an almost absurd conclusion. Our cities are dotted with banks that post the time in large electronic lights that flash minute by minute and sometimes second by second (there is one in Piccadilly Circus which actually tells the time in tenths of a second). From the seventeenth century on, the clock became a metaphor for the universe itself.[10]

Clearly, then, one can speak of a general "congruence" between science and capitalism in early modern Europe. The rise of linear time and mechanical thinking, the equating of time with money and the clock with the world order, were parts of the same transformation, and each part helped to reinforce the others. But can we make our case more strongly? Can we illustrate the interaction in terms of problems picked, methods used, solutions found, in the careers of individual scientists? In what follows, I shall attempt to demonstrate how these trends crystallized within the mind of Galileo, a figure so central to the Scientific Revolution. But our understanding of Galileo depends in part on our awareness of yet another aspect of the changes described above: the erosion of the barrier between the scholar and the craftsman which occurred in the sixteenth century. For many scientists, including Galileo, it was the availability of a new type of intellectual input which enabled their thoughts to take such novel directions.

Much has been made of the refusal of the College of Cardinals to look through Galileo's telescope, to see the moons of Jupiter and the craters on the surface of the moon. In fact, this refusal cannot be ascribed to simple obstinacy or fear of truth. In the context of the time, the use of a device crafted by artisans to solve a scientific (let alone theological) controversy was considered, especially in Italy, to be an incomprehensible scrambling of categories. These two activities, the pursuit of the truth and the manufacture of goods, were totally disparate, particularly in terms of the social class associated with each. Bacon's argument for a relationship between craft and cognition had as yet made little headway even in England, a country that, compared to Italy, had undergone an enormous acceleration in industrial production. Galileo, who studied projectile motion in the Venice arsenal, conducted scientific studies in what amounted to a workshop, and claimed to understand astronomy better by means of a manufactured device, was something of an anomaly in early seventeenth-century Italy. Where did such a person come from?

It was not until the late fifteenth century that the strong intellectual bias against craft activity, with its lower-class associations, began to break down. The crisis in the feudal economic system was accompanied by a historically unprecedented increase in the social mobility of the artisan class (including sailors and engineers).[11] At the same time, scholarly attacks on Aristotle (and they were not typical) drew ammunition from the history of technological progress, and in doing so lavished praise on the now exalted artisan, "who sought truth in nature, not in books."[12] The result—and the trickle which began ca. 1530 became a torrent by 1600—was a host of technical works published by artisans (very much an aberration in terms of class structure) and an increasing number of methodological critiques of Aristotelian-Scholastic science based on its complete passivity vis-à-vis nature. This new "mechanics literature," which was written in vernacular tongues, became popular among merchants and businessmen and was frequently reprinted. The breakthrough of artisans, craftsmen, engineers, and mariners into the ranks of publishing and scholarship, notes historian Paolo Rossi, "made possible that collaboration between scientists and technicians and that copenetration of technology and science which was at the root of the great scientific revolution of the seventeenth century."[13]

By and large, the artisan classes were simply asking that their work receive a hearing, not seeking a theory of knowledge based on technology; and those writers who did claim that technical activity constituted a mode of cognition (Bacon included) were at a loss as to what such a merger of theory and practice would look like. Yet the period 1550–1650, says Rossi, saw "continuous discussion, with an insistence that bordered on monotony, about a logic of invention. . . ."[14] Technology was hardly new in the sixteenth century, of course, but the level of its diffusion and the insistence on its being a mode of cognition *were* novel, and these events inevitably began to have an impact on scientists and thinkers. No longer restricted to such devices as catapults and water mills, technology became an essential aspect of the mode of production, and, as such, it began to play a corresponding role in human consciousness. Once technology and the economy became linked in the human mind, the mind started to think in mechanical terms, to see mechanism in nature—as Robert Hooke recognized. Thought processes themselves were becoming mechanico-mathematico-experimental, that is to say, "scientific." The merger of scholar and craftsman, geometry and technology, was now occurring within the individual human mind.

The change in attitude to artisanry on the part of some scholars also led to the rediscovery and sixteenth-century reprinting of a large number of classical technical works, including those of Euclid, Archimedes, Hero, Vitruvius, Apollonius, Diophantus, Pappus, and Aristarchus. Whereas much of previous mathematics had been conceived in terms of numerology, Pythagorean number mysticism, or even ordinary arithmetic, it was now increasingly possible to approach it from the point of view of an engineer. This development was to have an enormous influence on the work of Galileo, among others.

We have seen that the Galilean method incorporated a denial of teleological explanations (emphasis on how, rather than why); the formulation of physical processes in terms of "ideal types," which reality can be tested against by experiment; and the conviction that mathematical descriptions of motion and other physical processes are the guarantors of precision, and thus of truth. We saw too that Galileo had a very practical approach to such investigations (actually, an engineering approach), and that his method explicitly involved distancing himself from nature in order to grasp it more carefully—an approach that I have called nonparticipating consciousness. It is perhaps no surprise, then, that Galileo's particular intellectual outlook stemmed from influences originating outside of the traditional academic framework. Despite his various professorships, he was directly involved with precisely those facets of the technological tradition which were impinging upon certain scholars as a result of the collapse of the dichotomy between scholar and craftsman. Rossi correctly calls Galileo the premier representative of the scholarly and technological traditions, but it is the latter that should be emphasized.[15] With professorships at Pisa and Padua, and contact with popes, dukes, and the educated elite, Galileo was destined for an academic career; but in terms of orientation he did not fit comfortably into such a context. Galileo had direct contact with sailors, gunners, and artisans. Two of his mentors (or heroes), Niccolò Tartaglia and Giovanni Benedetti, had no university education whatever; another, Guido Ubaldo, studied mathematics on his own; and a fourth, Ostilio Ricci, was a professor at the Accademia del Disegno (School of Design) in Florence, a place that was turning out a new breed of artist-engineer. All four of these men stood at the forefront of the Renaissance revival of Archimedes, who had been as much an engineer as a mathematician. Tartaglia and Benedetti were also steeped in technical fieldwork. The former was the founder of the science of ballistics, his book *New Science* (1537) emerging out of problems he had encountered with the artillery at Verona in 1531; and Benedetti, an early Copernican who vigorously criticized Aristotle and held that bodies of unequal density fell with equal speed, served as court engineer at Parma and Turin. In short, Galileo was unique in the early seventeenth century. He was heir to the new mechanics, which had developed entirely outside the university; but significantly, he himself was located in an academic setting.

Although it is not possible, in this brief discussion, to elaborate in any greater detail on Galileo's intellectual antecedents, some comments on Tartaglia are in order because his works and style provide a major clue to Galileo's methodology. *New Science* was the earliest attempt to apply mathematics to projectiles, and it dealt extensively with the trajectories of cannonballs. Tartaglia was first to break with the Aristotelian notion of discontinuous trajectories, to state that the projectile path was curvilinear, and to demonstrate that the maximum range of a projectile occurred at a gun elevation of 45 degrees. Contradicting Aristotle, he claimed that the air resisted motion, rather than assisting it. Between the covers of a book on ballistics, then, Tartaglia advanced a theoretical analysis of motion. This same combination occurred in a book he wrote in 1551 on the raising of sunken vessels, a topic of obvious interest to a republic like Venice. To this study he appended his Italian translation of Archimedes' essay *On Bodies in Water*. Again, the text emerged not merely as a technical treatise, but as the first open challenge to Aristotle's law of falling bodies, for it used Archimedes' theory of buoyancy and surrounding media to argue against Aristotle's rigid distinction between up and down. Galileo was to follow in Tartaglia's footsteps, arguing that there was no natural upward motion; using Archimedes to overturn Aristotle; refining the mathematics of projectile motion; and intimately connecting, as Tartaglia had done in all his work, technical fieldwork with theoretical conclusions.

Galileo's involvement in technical problems was most intense during the so-called Paduan period (1592–1610) when he was engaged in his studies of motion. His own laboratory was like a workshop, where he manufactured mathematical apparatus. Galileo tutored privately on mechanics and engineering; did research on pumps, the regulation of rivers, and fortress construction; and brought out his first printed work, on the military compass, or "sector," as it was called. He also invented the "thermobaroscope," and took a strong interest in the field of engineering (now called materials science) which deals with the resistance of materials. Although Galileo made a distinction in his own mind between craft and theory, he broke with the prevailing view that saw them as totally unrelated. He was not just a scientist who also happened to be interested in technology, but rather used technology—both in spirit and method—as the source of theory. His last work, the *Two New Sciences*, opens with the following conversation between two imaginary interlocutors:

> *Salviati:* The constant activity which you Venetians display in your famous arsenal suggests to the studious mind a large field for investigation, especially that part of the work which involves mechanics; for in this department all types of instruments and machines are constantly being constructed by many artisans, among whom there must be some who, partly by inherited experience and partly by their own observations, have become highly expert and clever in explanation.

> *Sagredo:* You are quite right. Indeed, I myself, being curious by nature, frequently visit this place for the mere pleasure of observing the work of those who, on account of their superiority over other artisans, we call "first rank men." Conference with them has often helped me in the investigation of certain effects including not only those which are striking, but also those which are recondite and almost incredible.[16]

The book not only contains a discussion of projectile motion, but also includes a table of ranges for firing. Galileo makes much of the value of his theory to gunners, but as it turns out, they did much more for his science than he did for theirs.

How exactly did the technological tradition surface in Galileo's studies of motion? He not only agreed with the literature of this tradition, that construction is a mode of cognition, that manipulating nature is a key to knowing it, but he also showed precisely how this type of investigation should be carried out.

The analysis of projectile motion, of course, was derived from a practical military problem, and was, at the same time, a crucial blow to Aristotelian physics. Since Aristotle divided motion into two types, forced and natural, he concluded that projectile motion (see Figure 9) had to be discontinuous, that is, it had to consist of a forced motion (throwing the object into the air) and a natural one (the descent to earth):

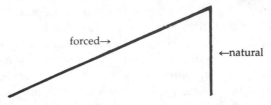

Figure 9. Aristotelian conception of projectile motion.

When people first hear about this theory, they often ask how intelligent men and women could have believed it, since all one has to do is *look* at a projectile to see that the above "curve" does not correspond to reality. In fact, the acceptance of Aristotle's theory is a good example of the gestalt principle of finding what you seek. Most readers probably have not watched a projectile very closely, and certainly few have plotted on a graph exactly where its apogee occurs and what then takes place. Furthermore, from the point of view of the thrower, a stone does seem to rise and then vertically drop. Finally, not until the end of the sixteenth century were cannon fired at long range, so such motion was not typically a part of the environment. As late as 1561 graphs in some textbooks were superimposed over a cannon, with the motion of the ball being shown as discontinuous (see Plate 1). In a world of qualitative science, the Aristotelian picture is roughly "true" in that it is one apparent aspect of projectile motion. Only with the rise of standing armies and the military concentration on ballistics was there any interest at all in a precise mathematical description of cannonball flight, which in any case is never really parabolic (see below) due to the effects of air resistance. We thus see how blurry, or complex, a simple "fact" can be: it seems to be shaped by what is being asked.

In any event, closer and closer scrutiny of projectiles made it more difficult to maintain the Aristotelian distinction between forced and natural motion. Since it is virtually impossible to map the points on a graph for an object actually thrown into the air, Galileo once again abstracted the essentials of the situation and

Plate 1. The Aristotelian theory of projectile motion, from Daniele Santbech, *Problematum Astronomicorum* (1561). Courtesy Ann Ronan Picture Library.

adapted them to laboratory conditions. Projectile motion, he reasoned, is a free-fall situation with a horizontal component. At the apogee of the curve, the object starts its downward descent due to the force of gravity, but it still retains some of the horizontal impetus originally imparted to it. The path would thus be smooth, not discontinuous, as Aristotle had maintained; and rather than abruptly falling to earth in a sheer vertical drop, the object would describe a curve, a combination ("resultant") of the vertical and horizontal components of motion. Galileo's experiments to ascertain this curve mathematically involved rolling a ball down an inclined plane that had a horizontal deflector at the bottom, and which was sitting on the edge of a table. The ball was released from different points along the plane, and thus in each trial struck the floor at a correspondingly different point. This generated a mass of data—really a collection of curves—which enabled Galileo, using his law of free-fall, to derive a mathematical description of these

63

curves as parabolic. In a nonresistant medium, he finally concluded, the trajectory of a projectile would be a perfect parabola.

The significance of this was not merely the mathematical description of a curve, but the challenge to Aristotelian physics. Not only did this weaken the distinction between forced and natural motion; it also called into question Aristotle's assertion that vacua could not exist (since projectile motion was supposedly maintained by displacing air rushing in to prevent a vacuum from forming), as well as the whole concept of immanent purpose contained in the Aristotelian doctrines of natural motion and natural place. Galileo's discovery of the independence of the horizontal and vertical components of motion, which is another aspect of the above investigation, led to his formulation of the composition and resolution of forces—what we now call vector mechanics. Here again, measurement, rather than any sort of purpose, is seen to lie at the heart of scientific explanation (if so it can be called). We see, then, that a military problem, which had been investigated by an engineer like Tartaglia, was converted into a controlled laboratory experiment to produce a mathematical expression, and then used to smash several fundamental tenets of the Aristotelian world view. Galileo's studies of ballistics not only refuted Aristotelian concepts; they were also beginning to delineate a new method for exploring reality.

All of Galileo's investigations served as vivid demonstrations of the relationship between theory and experiment which was slowly forming in the minds of a few European thinkers. They also vindicated the unproven assumption made by the technological literature of the sixteenth century: there *can* be a fundamental link between cognition and manipulation, between scientific explanation and mastery of the environment. The economic history sketched in the early pages of this chapter is thus much more than an interesting backdrop to these developments in the seemingly abstract realm of scientific thought. Cognition, reality, and the whole Western scientific method are integrally related to the rise of capitalism in early modern Europe.

We have talked in terms of a gestalt principle, of facts being plastic, "created" by theoretical constructs that are in turn linked to a socioeconomic context; and of the Scientific Revolution and its methodology as being part of a larger historical process. We are then brought face to face with an unsettling question: Is reality nothing more than a cultural artifact? Are Galileo's discoveries not the hard data of science, but simply the products of a world view that is a more or less localized phenomenon? If, as the foregoing analysis suggests, the answer is yes, we are cast adrift on a sea of radical relativism. There is then no Truth, but merely your truth, my truth, the truth of this time or that place. This is the implication of what is commonly called the sociology of knowledge. The distinction between knowledge and opinion, between science and ideology, crumbles, and what is right becomes a matter of majority rule, or "mob psychology."[17] Modern science, astrology, witchcraft, Aristotelianism, Marxism, whatever—all become equally true in the absence of objective knowledge and the concept of a fixed, underlying reality. Is there no way to protect ourselves from such a conclusion?

My answer is that radical relativism arises out of the peculiar attitude that modern science has adopted toward participating consciousness, which I discussed very briefly in the Introduction. It will be necessary, in the first place, then, to analyze the nature of participating consciousness in some detail. To do so, we must pursue the sociology of knowledge into a neglected chapter in the story of the Scientific Revolution: the world of the occult.

3

The Disenchantment of the World (1)

What appears a wonder is not a wonder.

—Simon Stevin

The phrase is Weber's: *die Entzauberung der Welt*. Schiller, a century earlier, had an equally telling expression for it: *die Entgötterung der Natur*, the "disgodding" of nature. The history of the West, according to both the sociologist and the poet, is the progressive removal of mind, or spirit, from phenomenal appearances.

The hallmark of modern consciousness is that it recognizes no element of mind in the so-called inert objects that surround us. The whole materialist position, in fact, assumes the existence of a world "out there" independent of human thought, which is "in here." And it also assumes that the earth, excepting certain slow evolutionary changes, has been roughly the same for millennia, while the people on that earth have regarded the unchanging phenomena around them in different ways at different times. According to modern science, the further back in time we go, the more erroneous are men's conceptions of the world. Our own knowledge, on this schema, is of course not perfect, but we are rapidly eliminating the few remaining errors that do exist, and shall gradually arrive at a fully accurate understanding of nature, free of animistic or metaphysical presuppositions. Modern consciousness thus regards the thinking of previous ages not simply as other legitimate forms of consciousness, but as misguided world views that we have happily outgrown. It holds that the men and women of those times *thought* they understood nature, but without our scientific sophistication their beliefs could not help but be childish and animistic. The "maturation" of the human intellect over the ages, particularly in this century, has (so the argument goes) almost completely corrected this accretion of superstition and muddled thinking.[1]

One of the goals of this chapter is to demonstrate that it is this attitude, rather than animism, which is misguided; and that this attitude stems, in part, from our inability to enter into the world view of premodern man. We have already established that modern science and capitalism were, historically, inextricably intertwined, and can appreciate that the perceptions and ideology of modern science are a part of large-scale social and economic developments. But because this scientific attitude is *our* consciousness, it is nearly impossible to abandon, even momentarily. Indeed, doing so is usually regarded as prima facie evidence for insanity. Nor does the recognition of the relativity our own consciousness serve, by itself, to place us at the center of a different consciousness. In short, it is very difficult to form a reliable impression of the consciousness of premodern society.

One thing that *is* certain about the history of Western consciousness, however, is that the world has, since roughly 2000 B.C., been progressively disenchanted, or "disgodded." Whether animism has any validity or not, there is no doubting its gradual elimination from Western thought. For reasons that remain obscure, two cultures in particular, the Jewish and the Greek, were responsible for the beginnings of this development. Although Judaism did possess a strong gnostic heritage (the cabala being its only survivor), the official rabbinical (later, talmudic) tradition was based precisely on the rooting out of animistic beliefs.[2] Yahweh is a jealous God: "Thou shalt have no other gods before me"; and throughout Jewish history, the injunction against totemism—worshipping "graven images"—has been central. The Old Testament is the story of the triumph of monotheism over Astarte, Baal, the golden calf, and the nature gods of neighboring "pagan" peoples. Here we see the first glimmerings of what I have called nonparticipating consciousness: knowledge is acquired by recognizing the *distance* between ourselves and nature. Ecstatic merger with nature is judged not merely as ignorance, but as idolatry. The Divinity is to be experienced within the human heart; He is definitely not immanent in nature. The rejection of participating consciousness, or what Owen Barfield calls "original participation," was the crux of the covenant between the Jews and Yahweh. It was precisely this contract that made the Jews "chosen" and gave them their unique historical mission.[3]

The Greek case is less easily summarized. At some point between the lifetime of Homer and that of Plato, a sharp break occurred in Greek epistemology so as to turn it away from original participation and contribute, out of very different motives, to the gradual disappearance of animism. It is difficult to conceive of a mentality that made virtually no distinction between subjective thought processes and what we call external phenomena, but it is likely that down to the time of the *Iliad* (ca. 900–850 B.C.) such was the case. The *Iliad* contains no words for internal states of mind. Given its contextual usage in this work, the Greek word *psyche*, for example, would have to be translated as "blood." In the *Odyssey*, however (a century or more later), *psyche* clearly means "soul." The separation of mind and body, subject and object, is discernible as a historical trend by the sixth century before Christ; and the poetic, or Homeric mentality, in which the individual is immersed in a sea of contradictory experiences and learns about the world through emotional identification with it (original participation), is precisely what Socrates and Plato intended to destroy. In the *Apology*, Socrates is aghast that artisans learn and pursue their craft by "sheer instinct," that is, by social osmosis and personal intuition. As Nietzsche pointed out, the phrase "sheer instinct," which in Socrates' mouth could only be an expression of contempt, epitomized the attitude of Greek rationalism toward any other mode of cognition. For this reason, he found Socrates (and indeed all of Western civilization) tragically inverted. The creative person, wrote Nietzsche, works by instinct and checks himself by reason; Socrates did just the reverse. And, Nietzsche continued, it was the Socratic form of rational knowledge which (despite Socrates' trial and sentencing) spread itself across the public face of Hellenism after his death.[4]

According to Eric Havelock, Plato regarded nonparticipating

71

consciousness, as exemplified by the Greek poetic tradition, as pathological.[5] Yet this tradition had been the principal mode of consciousness in Greece down to the fifth or sixth century before Christ, and during that period it served as the sole vehicle for learning and education. Poetry was an oral medium. It was recited before a large audience that memorized the verses in a state of autohypnosis. Plato used the term *mimesis,* or active emotional identification, to describe this submission to the spell of the performer, a process with physiological effects that were both relaxing and erotic, and that involved a total submergence of oneself into the other. Pre-Homeric Greek life, concludes Havelock, "was a life without self-examination, but as a manipulation of the resources of the unconscious in harmony with the conscious, it was unsurpassed."

Plato himself represented a relatively new tradition, one that sought to analyze and classify events rather than "merely" experience or imitate them. He spoke for the notion that subject was not object, and that the proper function of the former was to inspect and evaluate the latter. This perception could never take place if subject and object were merged in the act of knowing; or, to be more precise, if they never diverged to begin with. In the poetic tradition, the basic learning process was a sensual experience. In contrast, the Socratic dictum "know thyself" posited a deliberately nonsensual type of knowing.

Plato's work thus marks the canonization of the subject/object distinction in the West. Increasingly, the Greek began to see himself as an autonomous personality apart from his acts; as a separate consciousness rather than a series of moods. Poetry, to Plato, spoke of contradictory experiences, described a "many-aspect man" of inconsistent traits and perceptions. Plato's own psychological ideal was that of an individual organized around a center (ego), using his will to control his instinct and thereby unify his psyche. Reason thus becomes the essence of personality, and is characterized by distancing oneself from phenomena, maintaining one's identity. Poetry, *mimesis,* the whole Homeric tradition, on the other hand, involves identification with the actions of other people and things—the surrendering of identity. For Plato, only the abolition of this tradition could create the situation in which a subject perceives by confronting separate objects. Whereas the Jews saw participating consciousness as sin, Plato saw it as pathology, the archenemy of the intellect. At bottom, says Havelock,

72

Platonism "is an appeal to substitute a conceptual discourse for an imagistic one."[6]

Of course, Plato did not have his victory overnight. As Owen Barfield points out, original participation, knowledge via imagery rather than concepts, survived in the West down to the Scientific Revolution. Throughout the Middle Ages men and women continued to see the world primarily as a garment they wore rather than a collection of discrete objects they confronted. Yet the mimetic tradition was severely attenuated from Plato's time on, for some form of objectivity was now present; and it was chiefly the alchemical and magical tradition that attempted to demonstrate how limited this objectivity was.

The "Hermetic wisdom," as it has been called, was in effect dedicated to the notion that real knowledge occurred only via the union of subject and object, in a psychic-emotional identification with images rather than a purely intellectual examination of concepts. As indicated, this outlook had been the essential consciousness of Homeric and pre-Homeric Greece. In the following analysis of the Renaissance and medieval world views, then, it will be understood that premodern consciousness was located, mentally speaking, somewhere between pre-Homeric consciousness and the objective outlook of seventeenth-century Europe. With the Scientific Revolution, the considerable remnants of original participation were finally ousted, and this process constituted a significant episode in the history of Western consciousness.

The sixteenth century was an unusual period in European intellectual history, one that witnessed a vigorous revival, or resurfacing, of the occult sciences, which church Aristotelianism had successfully kept out of sight during the Middle Ages. Yet despite its vast differences from medieval Aristotelianism, the alchemical world view had in fact permeated medieval consciousness to a significant degree. Aristotle's doctrine of natural place and motion, for example, was part of the magical doctrine of sympathy, that like knows like; and the notion that the excitement of "homecoming" causes a body in free-fall to accelerate as it nears the earth is certainly an expression of participating consciousness. Furthermore, the highly repetitive and meditative nature of alchemical operations (grinding, distilling, and so on), which would induce altered states of consciousness through a prolonged narrowing of attention, was duplicated in hundreds of medieval craft techniques such as stained glass, weaving, calligraphy, metalworking, and the illumination of manuscripts. In general, medieval life and thought were significantly affected by animistic and Hermetic notions, and to some extent can be discussed as a unified consciousness.[7]

What were the common denominators of that consciousness? What did knowledge consist of, given the epistemological framework of sixteenth-century Europe? In a word, in the recognition of resemblance.[8] The world was seen as a vast assemblage of correspondences. All things have relationships with all other things, and these relations are ones of sympathy and antipathy. Men attract women, lodestones attract iron, oil repels water, and dogs repel cats. Things mingle and touch in an endless chain, or rope, vibrated (wrote Della Porta in *Natural Magic*) by the first cause, God. Things are also analogous to man in the famous alchemical concept of the microcosm and the macrocosm: the rocks of the earth are its bones, the rivers its veins, the forests its hair and the cicadas its dandruff. The world duplicates and reflects itself in an endless network of similarity and dissimilarity. It is a system of hieroglyphics, an open book "bristling with written signs."

How, then, does one know what goes with what? The key, as one might imagine, consists in deciphering those signs, and was appropriately termed the "doctrine of signatures." "Is it not true," wrote the sixteenth-century chemist Oswald Croll, "that all herbs, plants, trees and other things issuing from the bowels of the earth are so many magic books and signs?" Through the stars, the Mind of God impressed itself on the phenomenal world, and thus knowledge had the structure of divination, or augury. The word "divination" should be taken literally: finding the Divine, participating in the Mind that stands behind the appearances. Croll gives as one example the "fact" that walnuts prevent head ailments because the meat of the nut resembles the brain in appearance. Similarly, a man's face and hands must resemble the soul to which they are joined, a concept retained in palmistry even as it is practiced today, and in the common proverb (in many languages) that "the eyes are the windows of the soul."

One of the clearest expositions of the doctrine of signatures is found in the work of the great Renaissance magician Agrippa von Nettesheim, his *De Occulta Philosophia* of 1533.[9] In chapter 33 of this book he writes:

All Stars have their peculiar natures, properties, and conditions, the Seals and Characters whereof they produce, through their rays, even in these inferior things, viz., in elements, in stones, in plants, in animals, and their members; whence every natural thing receives, from a harmonious disposition and from its star shining upon it, some particular Seal, or character, stamped upon it; which Seal or character is the significator of that star, or harmonious disposition, containing in it a peculiar Virtue, differing from other virtues of the same matter, both generically, specifically, and numerically. Every thing, therefore, hath its character pressed upon it by its star for some particular effect, especially by that star which doth principally govern it.

Given this system of knowledge, modern distinctions between inner and outer, psychic and organic (or physical), do not exist. If you wish to promote love, says Agrippa, eat pigeons; to obtain courage, lions' hearts. A wanton woman, or charismatic man, possesses the same virtue as a lodestone, that of attraction.[10] Diamonds, on the other hand, weaken the lodestone, and topaz weakens lust. Everything thus bears the mark of the Creator, and knowledge, says Agrippa, consists of "a certain participation," a (sensuous) sharing in His Divinity. This is a world permeated with meaning, for it is according to these signatures that everything belongs, has a place. "There is nothing found in the whole world," he writes, "that hath not a spark of the virtue [of the world soul]." "Every thing hath its determinate and particular place in the exemplary world."

During his lifetime Agrippa was branded a charlatan and conjurer, and as we have noted, magic and Hermeticism were in continual conflict with the church. But this conflict, like the theory of knowledge that underlay it, was also one of resemblance, for the medieval church (as we shall discuss below) was steeped in magical practices and sacraments from which it derived its power on the local level. Consequently, it would tolerate no rivalry on this score.[11] The important point, however, is that all premodern knowledge had the same structure. As Michel Foucault tells us, divination "is not a rival form of knowledge; it is part of the main body of knowledge itself." Erudition and Hermeticism, Petrarch and Ficino, ultimately inhabited the same mental universe.

It is the collapse of this mental universe, beginning (if such a thing can be dated) in the late sixteenth century, that so radically marks off the medieval from the modern world; and nowhere is this more clearly portrayed than in Cervantes' epic, *Don Quixote*.[12] The Don's adventures are an attempt to decipher the world, to transform reality itself into a sign. His journey is a quest for resemblances in a society that has come to doubt their significance.

Hence, that society judges him to be mad, "quixotic." Where he sees the Shield of Mambrino, Sancho Panza can make out only a barber's basin; where (to take the most famous example) he perceives giants, Sancho sees only windmills. Hence the literal meaning of *paranoia*: like knowledge. The division of psychic and material, mind and body, symbolic and literal, has finally occurred. The madman perceives resemblances that do not exist, that are seen as not signifying anything at all. By 1600 he is "alienated in analogy," whereas four or five decades earlier he was the typical educated European. For the madman the crown makes the king, and Shakespeare captured the shift in the definition of reality in his line, "All hoods do not monks make." Given the meaninglessness of such associations, practices such as conjuring could no longer be regarded as effective. "I can call spirits from the vasty deep," says Glendower to Hotspur in *Henry IV, Part I*. "Why so can I, or so can any man," replies the latter; "But will they come when you do call for them?"

Hotspur's words are the first steps toward a relationship with the world with which we are very familiar. Glendower, on the other hand, sounds the last chords of a world largely lost to our imaginations; a world of resonance, resemblance, and incredible richness. Yet these chords may, even today, echo vaguely in our subconscious minds. Before turning to a more extended discussion of the collapse of original participation, then, it will be worth our while to stay with it a bit longer, and see if we cannot feel our way into this manner of thinking.

Participation is self and not-self identified at the moment of experience. The pre-Homeric Greek, the medieval Englishman (to a lesser extent, of course), and the present-day African tribesman know a thing precisely in the act of identification, and this identification is as much sensual as it is intellectual. It is a *totality* of experience: the "sensuous intellect," if the reader can imagine such a thing. We have so lost the ability to make this identification that we are left today with only two experiences that consist of participating consciousness: lust and anxiety. As I make love to my partner, as I immerse myself in her body, I become increasingly "lost." At the moment of orgasm, I *am* the act; there is no longer an "I" who experiences it. Panic has a similar momentum, for if sufficiently terrified I cannot separate myself from what is happening to me. In the psychotic (or mystic) episode, my skin has no boundary. I am out of my mind, I have become my environment. The essence of original participation is the *feeling*, the bodily perception, that there stands behind the phenomena a "represented" that is of the same nature as me—*mana*, God, the world spirit, and so on.[13] This notion, that subject and object, self and other, man and environment, are ultimately identical, is the holistic world view.

Of course, we sometimes experience participation in less intense forms, although sexual desire and panic remain the best examples. In truth—and we shall treat this in detail in Chapter 5—participation is the rule rather than the exception for modern man, although he is (unlike his premodern counterpart) largely unconscious of it. Thus as I wrote the first few pages of this chapter, down to this page, at least, I was so absorbed in what I was doing that I had no sense of myself at all. The same experience happens to me at a movie, a concert, or on a tennis court. Nevertheless, the consciousness of official culture dictates my "recognition" that I am not, and can never be, my experiences. Whereas my premodern counterpart felt, and saw, that he was his experiences—that his consciousness was not some special, independent consciousness—I classify my own participation as some form of "recreation," and see reality in terms of the inspection and evaluation Plato hoped men would achieve. I thus see myself as an island, whereas my medieval or ancient predecessor saw himself more like an embryo. And although there is no going back to the womb, we can at least appreciate how comforting and meaningful such a state of mind, and view of reality, truly was.

But was this view at all real? Weren't my predecessors simply living in the same world as I am, but somehow conceptualizing it differently (i.e., incorrectly)? Doesn't the subject/object dichotomy represent a distinct advance in human knowledge over this primitive, even orgiastic identification of self and other? These questions, which are all essentially asking the same thing, are the ones most crucial to the history of consciousness, and require closer scrutiny. For there are only two possibilities here. Either original participation, which was the basic mode of human cognition (despite the gradual attenuation of that mode) down to the late sixteenth century, was an elaborate self-deception; or original participation really did exist, was an actual fact.[14] We shall try to decide between these two alternatives by means of an analysis of the paradigm science of participation, alchemy.

If the standard history textbooks are to be believed, alchemy was the attempt to find a chemical substance that, when added to lead, transformed it into gold. Alternatively, it was the attempt to prepare a liquid, the *elixir vitae*, that would prolong human life indefinitely. Since neither of these goals is attainable, the entire alchemical enterprise is dismissed as a nonsensical episode (more than two thousand five hundred years) in the history of science, a venture that could be viewed as tragic were it not so silly in content. At most, modern science concedes that the alchemists did, in the pursuit of their spurious ends, discover as by-products various medicines and chemical substances that have some utilitarian value.

As is the case with all clichés, this one contains something of the truth. The quick production of the *lapis*, or philosopher's stone, whether in the form of gold or elixir, was certainly an irresistible goal for many alchemists, and the term "puffer" was used to denote the commercial opportunist and charlatan. "Of all men," wrote Agrippa, "chymists are the most perverse."[15] Yet a brief perusal of medieval and Renaissance alchemical plates, such as those collected by Carl Jung, is enough to convince us that such charlatanry was hardly the whole story to alchemy.[16] What could these strange images (see Plates 2–6) possibly mean? A green and red snake swallowing its tail; an "androgyne," or man-woman, joined at the waist with an eagle rising behind it and a pile of dead eagles at its feet; a green lion biting the sun, with blood (actually mercury) dripping from the resultant "wound"; a human skeleton perched on a black sun; the sun casting a long shadow behind the earth—these and other images are so fantastic as to defy comprehension. Surely, if all one wanted was health or wealth, there was no need for the painstaking preparation of such elaborately illustrated manuscripts. Mythopoeic artwork of this sort forces us to abandon the simplistic utilitarian interpretation of alchemy and try, instead, to chart the totally unfamiliar terrain of consciousness that this bizarre imagery represents.

It was the achievement of Carl Jung first to decipher the symbols of alchemy by means of clinical material from dream analysis, and then on this basis to formulate the argument that alchemy was, in essence, a map of the human unconscious. Central to Jungian psychology is the concept of "individuation," the process whereby a person discovers and evolves his Self, as opposed to his ego. The ego is a persona, a mask created and demanded by everyday social interaction, and, as such, it constitutes the center of our conscious life, our understanding of ourselves through the eyes of others.

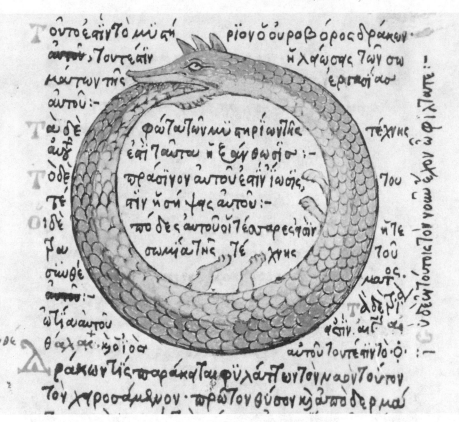

Plate 2. The Ourobouros, symbol of integration. Synosius, Ms. grec 2327, f.279, Phot. Bibl. nat. Paris

The Self, on the other hand, is our true center, our awareness of ourselves without outside interference, and it is developed by bringing the conscious and unconscious parts of our mind into harmony. Dream analysis is one way of achieving this harmony. We can unlock our dream symbols and then act on the messages of our dreams in waking life, which in turn begins to alter our dreams. But how to analyze our dreams? They are frequently cryptic, and so often violate causal sequence as to border on gibberish. But it is precisely here, Jung discovered, that alchemy can make a crucial contribution. In fact, it is by something like the doctrine of signatures that we are able to figure out what our dreams mean.[17]

The language of alchemy, as well as of dreams, follows a type of reasoning which I have termed "dialectical," as opposed to the

Plate 3. The alchemical androgyne. *Aurora consurgens*, Ms. Rh 172, Zentralbibliothek Zürich.

Plate 4. The green lion swallowing the sun. Arnold of Villanova, *Rosarium philosophorum* (1550), Ms. 394a, f.97, Kantonsbibliothek (Vadiana), St. Gallen.

Plate 5. *Sol niger:* the *nigredo;* from J. D. Mylius, *Philosophia reformata* (1622). Reproduced by C. G. Jung in *Gesammelte Werke,* publ.Walter-Verlag.

critical reason characteristic of rational, or scientific, thought.[18] As we saw earlier, Descartes regarded dreams as perverse because they violated the principle of noncontradiction. But this violation is not arbitrary; rather, it emerges from a paradigm of its own, one that could well be called alchemical. This paradigm has as a central tenet the notion that reality is paradoxical, that things and their opposites are closely related, that attachment and resistance have the same root. We know this on an intuitive level already, for we speak of love-hate relationships, recognize that what frightens us is most likely to liberate us, and become suspicious if someone accused of wrongdoing protests his or her innocence too hotly. In

82

short, a thing *can* both be and not be at the same time, and as Jung, Freud, and apparently the alchemists all understood, it usually is.

Within the context of the alchemical paradigm, it is critical reason that appears meaningless, and actually rather stupid, in its attempt to rob significant images of their meaning. Thus, in the example given in Chapter 1, if I dream that I am my father and that I am arguing with him, it is irrelevant that this is not logically or empirically possible. What *is* relevant is that I awake from the dream in a cold sweat and remain troubled for the rest of the day; that my psyche is in a state of civil war, torn between what I want for myself and what my (introjected) father wants for me. To the extent that this dilemma remains unresolved, I shall be fragmented, un-whole; and since (Jung believed) the drive for wholeness is inherent in the psyche, my unconscious will send out dream after

Plate 6. *The sun and his shadow complete the work,* from Michael Maier, *Scrutinium chymicum* (1687). Reproduced by C. G. Jung in *Gesammelte Werke,* publ. Walter-Verlag.

dream on this particular theme until I take steps to resolve the conflict. And because life is dialectical, so too will be my dream images. They will continue to violate the logical sequences of space and time, and to represent opposing concepts that, on closer examination, prove to be pretty much the same.

Jung's specific contribution, both to the history of alchemy and to depth psychology, was the discovery that patients with no previous knowledge of alchemy were having dreams that reproduced the imagery of alchemical texts with a bewildering similarity. In his famous essay "Individual Dream Symbolism in Relation to Alchemy," Jung recorded a series of one such patient's dreams and produced for nearly every dream a separate alchemical plate that duplicated the dream symbols in an unmistakable way.[19] Inasmuch as Jung claimed that others had produced a similar set of dream images while undergoing the individuation process, Jung was forced to conclude that this process was indeed inherent in the psyche and that the alchemists, without really knowing exactly what they were doing, had recorded the transformations of their own unconscious which they then projected onto the material world. The gold of which they spoke was thus not really gold, but a "golden" state of mind, the altered state of consciousness which overwhelms the person in an experience such as the Zen *satori* or the God-experience recorded by such Western mystics as Jacob Boehme (himself an alchemist), St. John of the Cross, or St. Theresa of Avila. Far from being some pseudo-science or proto-chemistry, then, alchemy was fully real—the last major synthetic iconography of the human unconscious in the West. Or, in Norman O. Brown's terms, "the last effort of Western man to produce a science based on an erotic sense of reality."[20]

Alchemy's rejection as a science, in Jung's view, coincided with the repression of the unconscious characteristic of the West since the Scientific Revolution—a repression that he saw as having tragic consequences in the modern era, including widespread mental illness and orgies of genocide and barbarism.[21] Thus Jung believed that the failure of each individual to confront his own psychic demons, the part of his personality he hated and feared (what Jung called the "shadow"), inevitably had disastrous consequences; and that the only hope, at least on the individual level, was to undertake the psychic journey that was in fact the essence of alchemy. In the cryptic words of the seventeenth-century alchemist and Rosicrucian, Michael Maier: *The sun and his shadow complete the work* (see Plate 6).[22] The creation of the Self lies not in repressing the unconscious, but in reintroducing it to the conscious mind.

Armed with this analysis, Jung found that the peculiar imagery represented in alchemical texts suddenly made sense. The "Ourobouros" of Plate 2, for example, a symbol that occurs (in one form or another) in almost every culture, represents the achievement of psychic integration, the unification of opposites. Green is the color of an early stage of the alchemical process, whereas red (the *rubedo*, as it was called) is that of a later one. Hence beginning and end, head and tail, alpha and omega, are united. The gold, or the Self, inherent from the first, is finally separated out. The world is the same, but the person has changed. As T. S. Eliot put it in "Little Gidding":

> We shall not cease from exploration
> And the end of all our exploring
> Will be to arrive where we started
> And know the place for the first time.

The dialectical nature of reality, which was embedded in the theory of resemblance, was captured in alchemy by pictures of androgynes (Plate 3), hermaphrodites, and brother-sister marriages or sexual unions. The conjunction of opposites occurs in the alchemical alembic, where lead is seen to be gold *in potentia*, where mercury is both liquid and metal, where what is volatile (represented by the rising eagle) becomes fixed, and what is fixed (the dead eagles at the bottom) volatile.

The danger of the work is the point of Plate 4, which depicts a green lion attempting to eat or swallow the sun. As already indicated, green is an early stage of the process, where the raw, vegetative force of the unconscious is released and the conscious mind feels itself in danger of being devoured. The alchemical slogan, "Do not use high-grade fires," is appropriate here. The cycle of sublimation and distillation is slow and infinitely tedious, as are all the alchemical operations, and any attempt to hasten the process will only prove abortive. The danger in tapping the unconscious is that one will get more than one bargained for; that the repressed unconscious will overwhelm the conscious as a hole is poked in the dike separating the two. This phenomenon is well known to many psychiatrists, as well as to many people who have studied yoga, meditation, or have experimented with psychedelic drugs ("high-grade fires").[23] The person in search of integration may be permanently scared off, or forced to undertake his or her search from the very beginning. At the very worst, the eruption of unconscious information can dismember the soul, result in psychosis.[24] The alchemical process is often summed up in the phrase *solve et coagula;* the persona is dissolved (on the psychic level) so as to enable the real Self to coagulate, or come together. But as R. D. Laing points out in *The Politics of Experience,* there is no guarantee that this Self will coagulate; indeed, such a result may be especially unlikely in a culture that is terrified of the unconscious and rushes to drug the individual back into what it defines as reality.[25] Even the relatively alchemical culture of the Middle Ages was keenly aware of such danger, as Plate 4 indicates; and it was part of the alchemical opus to "tame" the green lion, or "cut off his paws"—an act that (in material terms, from our point of view) consisted in touching sulfur with mercury or boiling it in acid for an entire day. If this taming were not carried out, the breakthrough of the unconscious, the dissolution of the ego, the collapse of the subject/object distinction, the sudden conviction that there is a Mind behind phenomenal appearances—this single, unified flash of light could catapult the practitioner into heaven or hell, depending on his or her makeup and the particular circumstances. Hence another crucial alchemical slogan: *Nonnulli perierunt in opere nostro*—"not a few have perished in our work."

Finally, Plate 5 represents the first phase of the work, the *nigredo,* in which the lead is dissolved and the solution becomes black. This is the "dark night of the soul," the point at which the persona has been dissolved and the Self has not yet appeared on the horizon. Hence the skeleton, the death of the ego, and the black sun (*sol niger*), representing acute depression. The "shadow" has now completely eclipsed the conscious ego. In *The Divided Self,* Laing quotes the writing of a schizophrenic patient who, with no previous knowledge of alchemy, uses the phrase "black sun" to describe her way of experiencing the world. But in dialectical fashion, lead contains the nugget of gold, and the skilled alchemist can bring about the transmutation by careful attention to his experiments. Hence the concluding line of Laing's book: "If one could go deep into the depth of the dark earth one would discover 'the bright gold,' or if one could get fathoms down one would discover 'the pearl at the bottom of the sea.' "[26]

Jung's analysis of alchemy is brilliant, and he produces provocative evidence that the alchemists were quite deliberate about the psychic aspect of their work. *Aurum nostrum non est aurum vulgi,* they write; "our gold is not the common [i.e., commercial] gold." *Tam ethice quam physice;* "as much moral as material." Or as one alchemist, Gerhard Dorn, candidly put it: "Transform yourselves into living philosophical stones!" Thus Jung was able to claim that what was "really" taking place in the alchemist's laboratory was the psychic process of self-realization, which was then projected onto the contents of the furnace or alembic. The alchemist *thought* he made gold, but of course he didn't; rather, he made some concoction that, due to his altered state of consciousness, he called "gold."

This hypothesis is a very attractive one, especially since we know that in the course of their work alchemists practiced a number of techniques that can produce these altered psychic states: meditation, fasting, yogic or "embryonic" breathing, and sometimes the chanting of mantras. These techniques have been practiced for millennia, especially in Asia, for the express purpose (in our terms) of breaking down the divide between the conscious and unconscious parts of the mind. They strip the person of mundane desires, enabling him to penetrate another dimension of reality; and as Western science is just beginning to discover, they are certainly efficacious in physiological terms, especially if we adopt the (to me, quite reasonable) position that soul is another name for what the body does. It is easy to assume that the psychic aspect is the reality, and the material aspect deluded or irrelevant.

Unfortunately, Jung's interpretation does not tell us anything about what the alchemist actually did with his pots and alembics. Instead, it extracts from his activity the portion that we find comprehensible, and discards the rest. Such an interpretation is, in short, the product of a modern scientific consciousness, assuming as it does that matter was forever the same, and that only mind (concepts of matter) changed. But the alchemical world view simply did not construct reality in our terms. The subject/object distinction was already blurry in the first place, and thus such an interpretation of reality makes no sense, for "projection" assumes a sharp dichotomy that the alchemist did not make. Obviously the alchemist was doing *something;* but the projection argument, although an improvement over the standard textbook version, still takes him less than seriously. The goal of magical practice was to become a skillful practitioner, not a self-realized being. The quotes from Dorn and other alchemists cited above are not typical, and they date from the sixteenth and seventeenth centuries, when the Scientific Revolution was relentlessly driving a wedge between matter and consciousness. For most of its history, alchemy had been perceived as an exact science, not a spiritual metaphor. If we succumb to Jung's formulation, we do so because of our inability to enter into a consciousness in which the technical and the divine were one, a consciousness in which finding a science of matter was equivalent to participating in God. Thus, Jung's formulation begs the question, for it is that very consciousness that we seek to penetrate. The very modernity of the projection concept precludes this possibility. The problem can only be solved (if at all) by trying to recreate the actual procedures of the discipline, and learning what the alchemist was doing in material terms.

Alchemy was first and foremost a craft, a "mystery" in medieval terminology, and all crafts, from the most ancient of times, were regarded as sacred activities. As Genesis tells us, the creation or modification of matter, the crux of all craftsmanship, is God's very first function. Metallurgy was intentionally compared to obstetrics: ores were seen to grow in the womb of the earth like embryos. The role of the miner or metalworker was to help nature accelerate its infinitely slow tempo by changing the modality of matter. But to do so was to meddle, to enter into sacred territory, and thus, down to the fifteenth century, the sinking of a new mine was accompanied by religious ceremonies, in which miners fasted, prayed, and observed a particular series of rites. In a similar fashion, the alchemical laboratory was seen as an artificial uterus in which the ore could complete its gestation in a relatively short time (compared to the action of the earth). Alchemy and mining shared the notion, then, that man could intervene in the cosmic rhythm, and the artisan, writes Mircea Eliade, was seen as "a connoisseur of secrets, a magician. . . ." For this reason, all crafts involved "some kind of initiation and [were] handed down by an occult tradition. He who 'makes' real things is he who *knows* the secrets of making them."[27]

From these ancient sources came the central notion of alchemy: that all metals are in the process of becoming gold, that they are gold *in potentia,* and that men can devise a set of procedures to accelerate their evolution. The practice of alchemy is thus not really playing God—though the notion is certainly latent in the Hermetic tradition—but is, to continue the obstetrical metaphor, a type of midwifery. The set of procedures came to be called the "spagyric art," the separating of the gross from the subtle in order to assist the evolution and obtain the gold that lay buried deep within the lead. "Copper is restless until it becomes gold," wrote the thirteenth-century mystic Meister Eckhart;[28] and although Eckhart may have had something more Jungian than metallurgical in mind, the alchemist, as we have stated repeatedly, made no such distinctions, but (in our terms) concentrated on his reagents and let nature (both human and inorganic) take its course.

What, then, were the procedures? Reading alchemical texts, the first thing one discovers is that there is very little unity of opinion on the subject. Transmutation consisted in the following set of operations: purification, solution, putrefaction, distillation, sublimation, calcination, and coagulation. However, the order and content of them is unclear, and not all alchemists employed all the techniques. Circumstances, especially the nature of the ores, always seemed to alter the methods. Hence what *is* agreed upon in terms of procedure is very general, consisting only of the basic outlines. Mercury is the dissolver, the active principle of things, and in fact had been used from the earliest times as a wash in gilding, to extract gold from other minerals. Sulfur (also called the green lion) is a coagulant, the creator of a new form. One must first perform the dissolution of the metal to the *materia prima* and then recrystallize, or coagulate, this formless substance. If done correctly, gold will be the result. *Solve et coagula* meant reduction to chaos—a watery solution, a primal state—followed by fixation into a new pattern.

In fact, the process was rarely this straightforward. The very delicacy of the procedure meant that it could be thrown off by the slightest mistake. Furthermore, it was central to the tradition that each student must learn this complex procedure by himself. There was no standardized recipe that could be handed on, but rather an elaborate practice that required a profound commitment. The variable factors were thus legion; failure rather than success was the rule. A number of intermediate steps, such as putrefaction, distillation, sublimation and calcination, normally had to be employed; and clearly, the terse formula *solve et coagula* expressed only an ideal.

Sometimes, the metal first had to be made to decay, or putrefy. The stink of this process came from hydrogen sulfide (the odor of rotten eggs), which was prepared and then passed through metallic solutions to obtain various colors (in the Middle Ages, colors

89

and odors were substantial entities, not secondary qualities). Or, an evaporable substance would have to be extracted from its mixture so as to obtain it in a pure state. Sulfur, in particular, was obtained in this way. Hence, the long and exacting process of sublimation that in turn necessitated the complementary process of distillation, or filtering. Finally, if a metal would not dissolve, calcination was employed to convert it to a soluble oxide so that the processes of solution and separation could be performed.[29]

That there are various psychoanalytic and religious correlates to these procedures is perhaps obvious. In a spiritual interpretation, all personalities (metals, ores) are potentially divine (golden), and are trying to reach their true nature, trying to transcend the weight of their past (lead). An old reality decays for me, I stink and feel rotten, but this change in matter is ultimately good, for it is a change in what matters. Old realities die, new things become my reality. The rigidity of my personality is dissolved, a new pattern is slowly allowed to coalesce. The ferocious desire for pattern itself is tamed, and I begin to look at my former pattern as just one possibility among many. I become less rigid, more tolerant. I see that all that really exists is fusability and creativity, which mercury represents. Mercury, or Hermes, the messenger of the gods, acts as "trickster" here, even though he is called "psychopomp," guide of the soul. As Freud realized, we have to be tricked into consciousness, see our true nature almost by accident, for example, through jokes or slips of the tongue. Mercury was also associated with glass, the vessel that enables one to see into it. The container of my problems is transparent: I come to see that my problems not only hold the solution, they *are* the solution. Thus R. D. Laing: "The Life I am trying to grasp is the me that is trying to grasp it."

The alchemist is thus like a miner, probing deeper and deeper veins of ore. One vein leads to another, there is no right answer. Life, and human personality, are inherently crazy, multifaceted; neurosis is the inability to tolerate this fact. The traditional model of the healthy soul demands that we impose an order or identity on all of these facets, but the alchemical tradition sees the result as an aborted metal that sulfur fixed too quickly. *Solve et coagula*, says the alchemist; abandon this prematurely congealed persona that forces you into predictable behavior and a programmed life of institutionalized insanity. If you would have real control over your life, says the tradition, abandon your artificial control, your "identity," the brittle ego that you desperately feel you must have for your survival. Real survival, the gold, consists in living according to the dictates of your own nature, and that cannot be achieved until the risk of psychic death is confronted directly. This, in the alchemical view, is the meaning of the Passion. When Christ said "I am the Way," he meant, "you yourself must go through my ordeal." No one else can confront your demons for you; no one else can give you your real Self.[30]

The conclusion seems unavoidable, then, that alchemy corresponds to a primal substrate of the unconscious, and both R. D. Laing and Jungian analyst John Perry have noted the identical imagery thrown up by the tortured psyche during the psychotic experience—imagery that is clearly alchemical in nature.[31] Still, the alchemist did not regard himself as a shaman or yogi, but as an expert on the nature of matter. Given the above description of laboratory procedures, what have we learned about the material aspect of the work? Essentially, nothing. That the alchemist was serious about his work, and the manufacture of gold, is beyond doubt. But what was he actually *doing* in his laboratory?

With this question we reach a total impasse. The literature of alchemy records that gold was in fact produced, and the testimony is not so easily dismissed. In one case, a transmutation was witnessed by Helvetius (Johann Friedrich Schweitzer), physician to the Prince of Orange, in 1666, and verified by a number of witnesses, including a Dutch assay master and a well-known silversmith. Spinoza himself got involved in the case, and reported the testimony without questioning it.[32] In the end, the answer to our question may depend only on whether or not one believes such a metallurgical transmutation is possible.

Nevertheless, I believe we can take this problem one step farther. Since the worlds constructed by participating and nonparticipating consciousness are not mutually translatable, the question, "What was the alchemist actually doing?" turns out to be something of a red herring when we examine what we mean by the word "actually." What we really mean is what *we* would be doing, or what a modern chemist would be doing, if we or he could be transported back in time and space to an alchemist's laboratory. But what was "actually" going on was what the *alchemist* was doing, not what we moderns, with our nonparticipating consciousness, would do if we could be transported back to the fourteenth century. Had we belonged to that era we would have possessed a participating consciousness and necessarily would have been doing what the alchemist was doing. Thus the question "What was the alchemist actually doing?" can have no meaningful answer in modern terms.

Let me put this another way. The world in which alchemy was practiced recognized no sharp distinctions between mental and material events. In such a context, there was no such thing as "symbolism" because everything (in our terms) was symbolic, that is, all material events and processes had psychic equivalents and representations. Thus alchemy was—from our viewpoint—a composite of different activities. It was the science of matter, the attempt to unravel nature's secrets; a set of procedures which were employed in mining, dyeing, glass manufacture, and the preparation of medicines; and simultaneously a type of yoga, a science of psychic transformation.[33] Because matter possessed consciousness, skill in transforming the former automatically meant that one was skilled in working with the latter—a tradition retained today only in fields such as art, poetry, or handicrafts, in which we tend (rightly or wrongly) to regard the ability to create things of great beauty as a reflection of the creator's personality. We say then, that the talent of the alchemist in his laboratory was dependent on his relationship with his own unconscious, but in putting it that way we indicate the limits of our understanding. "Unconscious," whether used by Jung or anyone else, is the language of the modern disembodied intellect. It was all one to the alchemist: there *was* no "unconscious." The modern mind cannot help but regard the occult sciences as a vast welter of confusion about the nature of the material world, since for the most part the modern mind does not entertain the notion that the consciousness with which the alchemist confronted matter was so different from its own. If the state of mind can at all be imagined, however, we can say that the alchemist did not *confront* matter; he *permeated* it.

It is thus doubtful that the alchemist could have described what he was doing to us, or to a modern chemist, transported back to the fourteenth century, even if he had wanted to. His was (again, from our point of view) partly a psychic discipline that no nonpsychic method (save neutron bombardment in a nuclear reactor) can possibly accomplish. The manufacture of gold was not a matter of replicating a material formula. Indeed, its manufacture was part of a much larger work, and our attempt to extract the material essence

92

from a holistic process reveals how contracted our own knowledge of the world has become. We cannot know the alchemical process of making gold until we know the "personality" of gold. We, here and now, have no real sympathetic identity with the process of becoming golden; we cannot fathom the relationship between becoming golden and making gold. The medieval alchemist, on the other hand, was completed by the process; the synthesis of the gold was his synthesis as well.

The only conclusion I can come to, then, is one that will probably strike most readers as radical in the extreme. The above analysis forces me to conclude that it is not merely the case that men conceived of matter as possessing mind in those days, but rather that in those days, matter *did* possess mind, "actually" did so. When the obvious objection is raised that the mechanical world view must be true, because we are in fact able to send a man to the moon or invent technologies that demonstrably work, I can only reply that the animistic world view, which lasted for millennia, was also fully efficacious to its believers. In other words, our ancestors constructed reality in a way that typically produced verifiable results, and this is why Jung's theory of projection is off the mark. If another break in consciousness of the same magnitude as that represented by the Scientific Revolution were to occur, those on the other side of that watershed might conclude that our epistemology somehow "projected" mechanism onto nature. But modern science, with the significant exception of quantum mechanics, does not regard the gestalt of matter/motion/experiment/quantification as a *metaphor* for reality; it regards it as the *touchstone* of reality. And if the criterion is going to be efficacy, we can only note that our own world view has pragmatic anomalies that are as extensive as those of either the magical or the Aristotelian world view. We are not, for example, able to explain psychokinesis, ESP, psychic healing, or a host of other "paranormal" phenomena by means of the current paradigm. There is no way, on a pragmatic basis, to make a judgment in terms of any epistemological superiority, and in fact, in terms of providing for a comprehensible world, original participation might even win out. Participation constitutes an insuperable historical barrier unless we consent to regenerate a dead evolutionary pattern—an act that would return us to a world view in which it would be meaningless to ask: Which epistemology is superior? Regenerating this pattern, we would, in some important sense, have fallen back through the rabbit hole whence we origi-

93

nally came. In such a world, the material transformation of lead to
gold may well occur, but we cannot know that now, nor can we
know it for the Middle Ages.

The delusion of modern thinking on alternative realities is rarely
exposed. Most historical and anthropological studies of witchcraft,
for example, never speculate that the massive number of witchcraft
trials during the sixteenth century might have been caused by
something more than mass hysteria. (Will our descendants, we
wonder, regard our involvement with science and technology as
mass hysteria, or more correctly realize that it was a way of life?)
The number of works that depict participating consciousness from
the inside, such as Chinua Achebe's description of Nigerian village
life in *Things Fall Apart*, is very small indeed; and I know of only
one writer who has managed both to enter that world and to articu-
late its epistemology in modern terms—Carlos Castaneda.[34] I shall
be discussing alternative realities in greater detail later on in this
book. For now, the reader should be aware of how stark the choice
really is. Either such realities were mass hallucinations that went
on for centuries, or they were indeed realities, although not com-
mensurable with our own. In his critique of Castaneda's work,
anthropologist Paul Riesman confronts the issue directly, though
the reader should note that Riesman hardly represents mainstream
thinking on the subject:

> Our social sciences [he writes] generally treat the culture and knowl-
> edge of other peoples as forms and structures necessary for human
> life that those people have developed and imposed upon a reality
> which we know—or at least our scientists know—better than they do.
> We can therefore study those forms in relation to "reality" and mea-
> sure how well or ill they are adapted to it. In their studies of the
> cultures of other people, even those anthropologists who sincerely
> love the people they study almost never think that they are learning
> something about the way the world really is. Rather, they conceive of
> themselves as finding out what other people's *conceptions* of the world
> are.[35]

In the case of the history of alchemy as well, or of premodern
thought in general, we have made precisely this mistake. We seek
to describe what the alchemist *thought* he was up to; we never
grasp that what he was "actually" doing was real. Moreover, we
rarely apply this methodology to our own methodology; we never
manage to see *our* culture and knowledge as "forms and structures

94

necessary for human life" as it exists in Western industrial societies.

The truth is that we can always find previous world views lacking if we judge them in our terms. The price paid, however, is that what we actually learn about them is severely limited before the inquiry even begins. Nonparticipating consciousness cannot "see" participating consciousness any more than Cartesian analysis can "see" artistic beauty. Perhaps Heraclitus put it best in the sixth century B.C. when he wrote, "What is divine escapes men's notice because of their incredulity."[36]

This brings us, finally, to the question of values, a question that is especially relevant because of the role of values in shaping our perceptions. Our purpose with respect to gold is not very different from that of King Midas. We seek to know how the alchemist "did it" because we see gold as a vehicle for obtaining other things. To the true alchemist, gold was the end, not the means. The manufacture of gold was the culmination of his own long spiritual evolution, and this was the reason for his silence. "The material aim of the alchemists," writes the historian Sherwood Taylor,

> the transmutation of metals, has now been realized by science, and the alchemical vessel is the uranium pile. Its success has had precisely the result that the alchemists feared and guarded against, the placing of gigantic power in the hands of those who have not been fitted by spiritual training to receive it. If science, philosophy, and religion had remained associated as they were in alchemy, we might not today be confronted with this fearful problem.[37]

By 1700, alchemy had been significantly discredited by the mechanical world view, or driven underground to become part of the ideology of so-called obscurantist groups: Rosicrucians, Freemasons, and others. In terms of making a claim on the dominant culture, its last great stand occurred during the English Civil War and Commonwealth period (1642–60), and its last great practitioner was Isaac Newton, though he wisely kept it a private matter.[38] Yet because alchemy (and all of the occult sciences) represents a map of the unconscious, because it apparently corresponds to a psychic substrate that is trans-historical, alchemy is still with us, both privately and publicly, and it is doubtful that dialectical reason can ever be completely extirpated. Privately it survives, as we have seen, in dreams, and also in psychosis.[39] Publicly it has but one surviving domain—the world of surrealist art. The express purpose of the Surrealist Movement in the first half of the twentieth century was to free men and women by liberating the images of the unconscious, by deliberately making such images conscious. There is, as a result, a peculiar visual link between alchemical plates, dreams, and surrealist art which seems to go deeper than appearances. All three use allegory and the incongruous juxtaposition of objects, and all three violate the principles of scientific causality and noncontradiction. Yet they do create a message by somehow managing to reflect, or evoke, certain familiar states of mind. These messages are intuitive, even numinous, rather than cognitive-rational, but we somehow "know" what they are saying. Their rules are those of premodern logic, of participating consciousness, of resemblance and "a secret affinity between certain images." "One cannot speak about mystery," wrote René Magritte; "one must be seized by it."[40] Hence the highly alchemical nature of a painting like *The Explanation* (Plate 7), in which a carrot and a bottle are both reasonably seen as distinct, and no less reasonably fused into a single object. Salvador Dali's *The Persistence of Memory* (Plate 8) has the same dreamlike quality, in which linear, mechanical time has started to wilt and run down in the arid desert of the twentieth century. Both of these paintings employ the same sort of logic and imagery that we observed in Plates 2–6.

We shall have to examine more closely what the public revival of alchemy in the twentieth century could possibly mean later on in this work. Our task now, however, is to try to solve the puzzle of why it was ever lost in the first place. Although we may have succeeded in immersing ourselves in that world view, we have not yet addressed the question of how modern science managed to refute it. The holistic framework of the occult sciences lasted for millennia, but it took Western Europe a mere two hundred years—roughly between 1500 and 1700—to break it apart, revealing that the Hermetic tradition was, despite its long tenure, rather fragile.

The problem lay in the tradition's (from our viewpoint) inherently dualistic nature. Magic was at once spiritual and manipulative, or, in D. P. Walker's terminology, subjective and transitive.[41] Each of the occult sciences, including alchemy, astrology, and the cabala, aimed at both the acquisition of practical, mundane objectives, and union with the Divinity. There was always a tension between these two goals (which is not the same thing as an an-

Plate 7. René Magritte, *The Explanation* (1952). Copyright © by A.D.A.G.P., Paris, 1981.

Plate 8. Salvador Dali, *The Persistence of Memory* (1931), oil on canvas, 9½" × 13". Collection, The Museum of Modern Art, New York.

tagonism) because they constituted a rather delicate ecological framework. If, for example, I am acting as a "midwife" to nature, accelerating its tempo in altering the nature of matter, it is clear that I am interfering in its natural rhythm. Any type of human action upon the environment can be seen in these terms. But the point is that the interference was always consciously acknowledged. It was sanctified through ritual, lest the earth strike back against man for this incursion into its womb. This interference was performed in the context of a mentality, and an economy (steady-state), that sought *harmony* with nature, and in which the notion of mastery of nature would have been regarded as a contradiction in terms. Nevertheless, the distinction ultimately involved a difference of degree rather than kind, for at what point in our acceleration of nature's tempo can we be said to have crossed the line from mid-wifery to induced birth, or even abortion? What degree of inter-ference tips the balance from harmony to attempted mastery? In a feudal context of subsistence economy and only moderately dif-fused technology, in a religious context that regarded nature as alive and our relationship to it as one of participation, it was very difficult for such a question to arise, and in this sense the alchemi-cal tradition was not all that fragile. But with the social and eco-nomic changes wrought in the course of the sixteenth and seven-teenth centuries, the sacred and the manipulative were split down the middle. The latter could easily survive in a context of profit, expanding technology, and secular salvation; indeed, that was what the manipulative aspect was all about, severed from its reli-gious basis. Thus Eliade rightly calls modern science the secular version of the alchemist's dream, for latent within the dream is "the pathetic programme of the industrial societies whose aim is the total transmutation of Nature, its transformation into 'energy.' "[42] The sacred aspect of the art became, for the dominant culture, ineffective and ultimately meaningless. In other words, the domination of nature always lurked as a possibility within the Hermetic tradition, but was not seen as separable from its esoteric framework until the Renaissance. In that eventual separation lay the world view of modernity: the technological, or the *zweckra-tional*, as a logos.

What is perhaps remarkable, from the modern point of view, is that magic could actually have served as a matrix for the Scientific Revolution. As explained in Chapter 2, technology had no theoret-ical or ideological basis, at least not until Francis Bacon. Even down to the time of Leonardo da Vinci, machines tended to be seen as toys, whereas the concept of force was linked to the Hermetic theme of universal animation.[43] Technology, in short, could not be a rival to Aristotelianism because it was not a *philosophy* about how the universe worked. Magic was. Of course, there were many types of magic and many magical philosophies, but all of them, in sharp contrast to church Aristotelianism, urged the practitioner to operate on nature, to alter it, not to remain passive. In this sense, then, the ascendancy of magical doctrines and techniques in the sixteenth century was fully congruent with the early phases of capitalism, and Keith Thomas has recorded (for England, at least) how extensive and intense occult activity was during this time.[44]

The idea of dominating nature arose from the magical tradition, perhaps the first explicit statement of the notion occurring in a work by Francesco Giorgio in 1525 (*De Harmonia Mundi*), which is not about technology, but—of all things—numerology. This art, he says, will confer upon man as regards his environment *vis operandi et dominandi*, "the power of operating and dominating." We should not be surprised that, in the sixteenth century, this concept was easily extended from numerology to accounting and engineering.

Numerology provides a very instructive example, in fact, of the split between the esoteric and exoteric traditions of the occult sciences. At the heart of the cabala, for example, lay the notion of a "dialing code." In the Hebrew alphabet, letters are also numbers, and hence an equivalence can be established between totally unrelated words based on the fact that they "add up" to the same amount. The right combination, it was believed, would put the adept in contact with God. Pico della Mirandola, for example, was fascinated by the mystical ecstasy brought on by number meditation, a trance in which communication with the Divinity was said to occur (the meditation could, of course, produce such ecstasy if the activity narrowed one's attention in a yogic fashion).[45] At the same time, similar techniques formed the basis of a practical cabala that the adept might use to obtain love, wealth, influence, and so on.

Under the pressure of the technical and economic changes of the sixteenth century, pursuit of God or world harmony began to seem increasingly quaint, and emphasis on the practical or exoteric tradition resulted in a purely representational use of the Hebrew alphabet. We can see this shift in books published only a decade apart by Robert Fludd and Joseph Solomon Delmedigo. In Plate 9, Fludd's illustration of the Ptolemaic universe (1619), the Hebrew letters signify the "spiritual intelligences" that rule each of the twenty-two spheres, from the World Mind ("Mens") down to the sphere of the earth. (This same type of labeling also occurs in cabalistic illustrations of the human body, where Hebrew letters serve to identify the spiritual intelligences that govern each particular part.) Fludd was a major proponent of the view that the Hebrew letters in the diagram corresponded to something real: they concretely identified the ruling archetypes of each region, and this information could be plugged into certain types of cabalistic "equations" to generate significant results for the practitioner. It was hardly a problem that the letters did not correspond to anything material or substantial in nature.

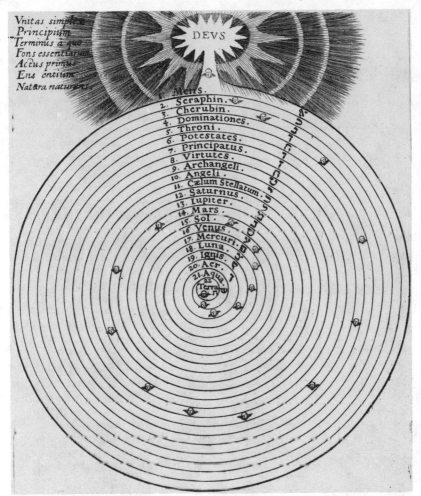

Vnitas simplice
Principium
Terminus a quo
Fons essentiarum
Actus primus
Ens entitum
Natura naturans

DEVS

Mens.
2. Seraphin.
3. Cherubin.
4. Dominationes.
5. Throni.
6. Potestates.
7. Principatus.
8. Virtutes.
9. Archangeli.
10. Angeli.
11. Cælum Stellatum.
12. Saturnus.
13. Iupiter.
14. Mars.
15. Sol.
16. Venus.
17. Mercuri.
18. Luna.
19. Ignis.
20. Aer.
21. Aqua.
22. Terra.

Plate 9. The Ptolemaic universe according to Robert Fludd, 1619. Courtesy, The Bancroft Library, University of California, Berkeley.

A very different use of the Hebrew alphabet is depicted in Plate 10, an engineering sketch from Joseph Solomon Delmedigo's book *Elim* (1629). Here, the letters are used to label a set of gears in a diagram illustrating how power can be multiplied so that, in Archimedean fashion, an individual with a place to stand can move a large section of the earth. It is no accident that Rabbi Delmedigo had been a student of Galileo at Padua, that he was an ardent Copernican, the first Jewish scholar to employ logarithms, and ultimately a leading popularizer of scientific knowledge. Yet the

101

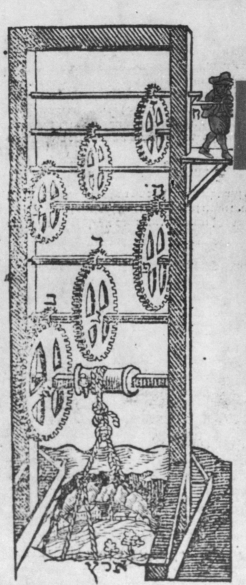

Plate 10. Engineering illustration from *Elim,* by Joseph Solomon Delmedigo, 1629.

labels have a still more complex significance. *Elim* means "powers" or "forces," and its implication can be both sacred and secular. Thus Jehovah is addressed as *El* in Hebrew liturgy; and more generally, an *el* can be a power that carries the essence ("spiritual intelligence") of God. But *el* can signify a purely material force as well, such as the power developed by a gear train. This ambiguity is reflected in the book itself, which deals with both religious and scientific matters, and in the author's attitude toward the cabala—an attitude that was so ambiguous that present-day Jewish scholars remain uncertain whether Delmedigo was a critic or a proponent. For a time, then, disparate concepts of number could exist side by side, even within a single mind, but ultimately, the esoteric tradition was unable to sustain itself. Under the pressures of a new economy, the spiritual aspect of the cabala, along with the evocative power of the spoken Hebrew word, became increasingly irrelevant. It was not that the cabala was "wrong," but that technology and mercantile capital had little use for religious mathematics.[46]

A similar transition occurred in all of the occult sciences during the sixteenth and early seventeenth centuries, with the possible exception of witchcraft, which was (to my knowledge) purely transitive and without a subjective, or self-transforming, aspect. What science accomplished (or rather, what became science) was the adoption of the epistemological framework, indeed the whole ideology, of the exoteric tradition. All of the "natural magicians" of the sixteenth century, such as Agrippa, Della Porta, Campanella, John Dee, and Paracelsus, right down to Francis Bacon, drew on both the technological and Hermetic traditions for the phrase "evoking the powers of nature." Both traditions began to fuse at this time and form the basis of modern scientific experimentation. Both were *active* ways of addressing reality, constituting a sharp contrast to the static nature of Greek science and the frozen verbalism of medieval Scholastic disputation. The identity of knowledge and construction which we discussed in Chapter 2, the "making that is itself a knowledge," which received its clearest expression in the work of Bacon, was derived from the numerous writings on magic and alchemy which appeared in Europe during the sixteenth century.[47] Della Porta candidly termed magic the "practical part of natural science," and such men as Dee, Campanella, and Agrippa tended to blur (from our point of view) control of the environment by means of the art of navigation with control of the environment by means of astrology.[48] Prior to and during the English Civil War, remarked John Aubrey in *Brief Lives*, "astrologer, mathematician, and conjurer were accounted the same things."[49] It was only after magic had provided technology with a methodological program that the latter was in a position to reject the former. But it was more in the fusion of the two, than in their subsequent separation, that the esoteric tradition was lost.

Examples of this sort can easily be multiplied. The esoteric tradition in astrology, for example, as represented by the Florentine scholar Marsilio Ficino (1433–99), who translated the entire Hermetic corpus for Cosimo de Medici between 1462 and 1484, sought to condition the body and spirit through music or incantation in order to alter the personality ("receive the celestial influence"). Bacon himself approved of this aspect of the art, calling it "astrologia sana," and D. P. Walker has in effect said the same thing when he calls Ficino's system "astrological psychotherapy."[50] But the ultimate legacy of the tradition, even among present-day astrologers who consider themselves serious students, is for the most part manipulative and this-worldly, and the horoscope column in the daily newspapers represents the pathetic outcome of what was once a magnificent edifice of dialectical thought.

In the case of alchemy, the causes of the exoteric-esoteric split were once again technological, particularly because of alchemy's age-old relationship to mining, metallurgy, and numerous crafts and manufactures. The sixteenth century saw the emergence of a coterie of artisans who denounced the alchemists, this attitude being most clearly expressed in works such as Biringuccio's *Pirotechnia* and Agricola's *De Re Metallica*.[51] The split is at the same time a response to changing economic relationships, in particular, the collapse of the guilds. An increasingly laissez-faire economy challenged both the feudal notion of maintaining secrecy about a craft's techniques and the oral tradition that had been the basis of initiation into these "mysteries." Pressure to reveal these secrets, to make them accessible to all by way of Gutenberg's movable type, led to the publication of craft handbooks (like those of Biringuccio and Agricola) which provided detailed accounts of processes and illustrations of guild practices (see Plate 11). These works, the appearance of which would have been viewed with horror in the Middle Ages, now served the interests of a large and amorphous social class. Craft processes themselves had become commodities; and secrecy, revealed knowledge, and microcosm/macrocosm analogies were seen as superfluous and even inimical

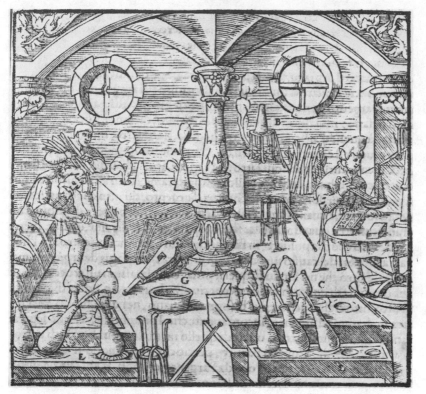

Plate 11. Separating gold from silver, from *De Re Metallica* (1556). Courtesy, The Bancroft Library, University of California, Berkeley.

by an artisanry that was increasingly caught up in a market economy. Thus, when the surgeon Ambroise Paré (1510–90) was accused of having betrayed guild secrets, he felt confident in replying that he was not of those men who "make a cabala of art."[52] Indeed, the whole notion of scientific organization which was trumpeted by Bacon in the *New Atlantis* was completely incompatible with the medieval ideal of deliberate secretiveness.

The ideology of this attack was heavily linguistic in nature. Once the idea of an inner psychic landscape (in our terms), or original participation, was partly lost, technology was able to judge the alchemical tradition from the point of view of technical clarity and precision and, of course, find it sorely lacking. As we have seen, the language of alchemy is dreamlike, symbolic, imagistic, but this world of resemblance was disintegrating. Carrots were not bottles, lions no longer devoured the sun, androgynes were inventions in the same category as unicorns. Cryptic phrases such as "the sun and his shadow complete the work" did not glaze pots or extract tin, and names for substances such as "butter of antimony" or "flowers of arsenic" (which, however, lasted down to the late eighteenth century) were now seen as cumbersome and inefficient. The whole alchemical imagery of things being themselves and their opposites, or possessing inherent ambiguity, was now regarded as stupid, incomprehensible, an obstacle to be rooted out. Biringuccio, Bacon, Agricola, Lazarus Ercker, and many others deliberately set themselves against the tradition of wonder at nature, of credulity about fabulous beasts and plants and stones—a tradition that had characterized medieval literature from Pliny to Agrippa. The notion of *satsang* still present in esoteric disciplines like Zen and yoga, that the truth is miraculously communicable through a relationship with a teacher, was an anathema to these men, who correctly saw that the attempted domination of nature depended on cognitive clarity. The collapse of an ecological, or holistic, orientation toward nature went hand in hand with the rejection of dialectical reason.[53]

The second factor contributing to the split between the esoteric and exoteric traditions was organized religion, both Catholic and Protestant. It was the very intimacy between magic and Catholicism which led to an exaggerated emphasis on alchemy's esoteric aspects (indeed, prior to this, alchemy was not seen as having "aspects"), an emphasis that served to sharpen the distinction between the esoteric tradition and the growing body of technological studies which were rejecting that tradition in the first place. This same intimacy also left magic extremely vulnerable to Protestant rationalism, both during and after the Reformation.

According to Keith Thomas, the church was quite heavily involved in magical practices on the local level during the Middle Ages. Indeed, without the network of rituals and sacraments it is doubtful that the church could have had the leverage that it did. The liturgy of the time included rituals for blessing houses, tools, crops, and people setting out on journeys; rituals to insure fertility; and rituals of exorcism. In the popular mind, the priest had special powers, and a whole range of beliefs, or superstitions, had grown up around the ceremony of the Mass. Thus the wafer was seen as having the power to cure the blind, and it could also be crushed and sprinkled in the garden to discourage caterpillars. At

106

the same time, the church deliberately blurred the distinction between prayers, which were appeals for supernatural help, and the tools of magic, such as charms or spells, which were supposed to work automatically. The church recommended the use of prayers when gathering medicinal herbs; and the repetition of *ave marias* or *pater nosters* fostered the notion that these Latin "incantations" had a mechanical efficacy. All in all, despite the church's opposition to magic on the official level, it appeared to the populace "as a vast reservoir of magical power, capable of being deployed for a variety of secular purposes."[54]

As for alchemy, its relationship to the church, at least during the Middle Ages, was practically heretical, for it occasionally claimed to provide the inner content of Christianity which it felt no organized religion could supply.[55] Thus it argued, every so often, for an analogy between Christ and the alchemical work itself, the so-called lapis-Christ parallel. This analogy and the claim of material transformation resulted in several encyclicals and papal bulls against the art, but as the social structure of the church began to crumble in the fifteenth century, alchemy and religion became intertwined in a most unusual way. In particular, the soteriological (salvationist) aspect of the art began to receive more attention even as the "puffers" and charlatans were subject to increasing attack. This development was, in fact, another facet of the esoteric-exoteric split. Sir George Ripley (1415–90), canon of Bridlington and an alchemist as well, frankly stated that the purpose of alchemy was the union of the soul with the body. By the sixteenth century, the church had drawn up a document establishing correspondences between the various alchemical processes and church sacraments. Hence putrefaction was extreme unction; distillation, ordination; calcination, repentance; coagulation, marriage; solution, baptism; sublimation, confirmation; and of course, transmutation, the Mass.[56] We might infer from these correspondences that the collapse of church magic under the pressure of heretical sects, and later, the Protestant Reformation, led to an overemphasis on the religious dimension of alchemy. This, in addition to the attack being mounted by the growing technological literature, ultimately served to split it off from the exoteric tradition.

It was during the Renaissance that the soteriological aspect of alchemy was pushed to its extreme, becoming, says Jung, "an undercurrent of the Christianity that ruled on the surface." In addition to the lapis-Christ parallel, some texts referred to mercury as the Virgin Mary, and the spirit of mercury as the Holy Ghost. Sir George Ripley constantly intermingled Christian and alchemical symbols in a way that turned into an unwitting parody of Catholicism. In one of his sketches, for example, the green lion lies bleeding in the lap of the virgin, an obvious caricature of the Pieta.[57] The Christian attitude toward alchemy at this time is also revealed in the choice of animals used as metaphors for Hermes, which were the same as had been used for Christ in patristic literature: dragon, fish, unicorn, eagle, lion, and snake. Transubstantiation was seen as, in essence, an alchemical process. Ripley and others praised the making of the stone as the Second Coming which, Jung notes, "sound[s] very queer indeed in the mouth of a medieval ecclesiastic." Indeed, what we see is an unwitting distortion of Christianity, an apotheosis that was at the same time a melting down. The medieval Christian synthesis was thus recast in alchemical terms, and this tendency reached its climax at the end of the sixteenth century with the rise of the Rosicrucians, a semisecret, occult brotherhood that still exists today.

By the end of the sixteenth century, the intimate relationship between magic and the church had become such an obvious target for the Reformation that magical practices of all kinds began to draw fire from Catholic as well as Protestant quarters. The story is rather complicated, because Catholic-Protestant relations themselves were very complex, and the attack on magic was part of an internecine cross fire that is not easily unraveled. Catholic opposition to magic was facilitated by a Protestant commitment to the Hermetic tradition on the part of those who, suggests Jung, saw that tradition (perhaps unconsciously) as a way of remaining Catholic. Thus toward the end of the sixteenth century in Germany, a group of occult practitioners began to argue openly for Hermeticism as being the path to divine illumination, explicitly stating the lapis-Christ parallel.[58] This group began to have an impact on Lutheran circles, and to rally behind those Protestant forces that could offer it protection from the long arm of the Inquisition. The movement thus acquired a political tinge, which emerged in anonymous manifestoes of 1614–15 defending Rosicrucianism and the occult sciences.

Europe soon found itself swept up in a frenzy over Rosicrucianism and its heretical implications. Orthodox religion was convinced of the existence of something approaching a world-wide conspiracy, a charge explicitly denied by the alchemist Michael

Maier in his *Laws of the Fraternity of the Rosie Cross* (Latin edition 1618)—a book that nevertheless affirmed the existence of a secret brotherhood of enlightened mystics dedicated to the improvement of mankind. Two years prior to this, the English physician and alchemist Robert Fludd published his own defense of the brotherhood (*Apologia Compendaria Fraternitatem de Rosea Cruce*) which he followed with a series of volumes from 1617 to 1621. Fludd argued for the inner content of the occult sciences, an alchemical interpretation of the Bible (e.g., seeing the creation as a divine chemical separation), and the view of nature as one vast alchemical process.

Of course, the emergence of a fraternity of alchemists arguing in support of alchemy, as well as of publications defending this group, probably reflected not the strength of the tradition but the fact that it was dying. As frightening as a defense of religious alchemy was to the church, it is clear from hindsight that it came about, in part, as an attempt on the part of some to maintain and preserve what they regarded as the genuine spiritual content of Catholicism. In the context of the times, however, alchemy's claim to provide the only true salvation could not be regarded as anything but pernicious heresy. Thus in 1623, a proclamation appeared in Paris announcing the arrival of the brotherhood, which declared that it would remain invisible but would lead people onto the true path. The following year, an open meeting held to defend alchemical theses was dispersed by order of Parlement, and its leading spokesman (one Estienne de Clave) arrested. It was in such a context that the Minorite friar Marin Mersenne set out to save both church and state, as well as philosophy itself, from this dangerous turn of events. This attack so snowballed, enlisting as it did the finest minds of Europe, that it has rightly been regarded as the death knell of animism in the West. It involved not merely a widespread rejection of esoteric alchemy, but possibly the first clear enunciation of both the fact-value distinction and the positivist conception of science.

As a man deeply interested in religion and natural philosophy, Mersenne was alarmed not only by the Rosicrucian phenomenon but also by the fact that the growing aversion of scholars to Aristotelianism had led them to Hermeticism, which offered a more active and experimental approach to nature. He saw that it would be necessary not only to refute Fludd, but to work out a Christianized version of Aristotelian rationalism which would simultaneously facilitate a more dynamic approach to the natural world.

In lengthy works written and published over the period 1623–25, Mersenne denounced Fludd as an "evil magician" and attacked alchemy as an attempt to offer salvation without faith, that is, to set itself up a counter-church. By attributing power to matter itself, the Hermetic tradition had denied the power of God, Who should rightly be seen as Governor of the world, not immanent in it. Instead of advocating the abolition of exoteric alchemy, however, Mersenne proposed something that was ultimately far more effective in this regard: that the state should establish alchemical academies to police the field of charlatans. These academies would clean up the language of alchemy, substituting a clear terminology based on observed chemical operations. They would also avoid all discussion of religion and philosophy. He proposed, in effect, the deliberate divorce of fact from value which would soon become the distinguishing hallmark of modern science.

In the course of his attack on Fludd, Mersenne enlisted the aid of his fellow Minorite Pierre Gassendi. A professor at Aix-en-Provence, Gassendi moved to Paris in 1624, eventually (through the influence of Cardinal Richelieu) becoming Provost of the Cathedral of Digne and Professor of Mathematics at the Collège Royale. His attack on Fludd was, like Mersenne's, religious, accusing the Englishman of trying to make alchemy "the sole religion of mankind"; but it was a scientific critique as well, arguing that Fludd's central notions could not be empirically demonstrated. There was no way, for example, to prove that all human souls contained a part of God, or that a world soul actually existed. Gassendi's attack on Fludd may have been, in effect, the earliest statement of scientific positivism. This equating of the measurable with the real was another version of the public stance Newton adopted when the concept of gravity was challenged as an occult cause.

Gassendi's attack, however, was much more than a critique. In the course of the 1630s he elaborated a world view of matter and motion that, despite its differences from the ideas of Hobbes and Descartes, amounted to a billiard-ball conception of the universe. Change was external, occurring through physical causation, rather than through the internal (dialectical) principles posited by the alchemists. All we can know, he argued, are appearances, not things in themselves. Matter, as well as the earth, is effectively dead; and God is not a world soul, but a world director.[59]

The similarities that the reader may have noted between Cartesian physics and the views of Mersenne and Gassendi are not accidental. Descartes was also close with Mersenne, moving to Paris in 1623 and contributing to the common effort of providing a Christianized atomism that would preserve religious and political stability. In the *Principles of Philosophy*, the world spirit of the al-chemists had become a world mechanism (ether moving in vor-tices), with mind expunged from matter and God relegated to the periphery. The destruction of participating consciousness, and the role of God as external director, were hardly unwitting features of the system. Both provided "scientific" sanctions against indepen-dent religious or political thought. As Descartes wrote Mersenne in 1630, "God sets up mathematical laws in nature, as a king sets up laws in his kingdom."

The collapse of alchemy was the result, not merely of learned publications, but of the very organization of science. Mersenne's monastic cell became the virtual nerve center of European science. He conducted weekly meetings and a vast correspondence with scientists in every country, introducing their works to each other and to the educated public. Proponents of mechanism, such as Galileo, were translated or explicated. Contacts were made with men who would later be key figures in the Royal Society of Lon-don, and these ties were strengthened when a number of them went into exile in Paris during the Civil War. Walter Charleton introduced Gassendi's ideas to England in 1654, and soon thereaf-ter Robert Boyle began a series of publications attacking alchemy and arguing for the mechanical world view, which, he tried to show by experiment, conformed to actual experience. Alchemical doctrines were "chemicalized" by a process of linguistic clarifica-tion and translation into strictly exoteric terms. The mechanical philosophy, and the divorce of fact from value, were built right into the guidelines of the Royal Society.

After Mersenne's death, Gassendi presided over the weekly meetings, which now took place at the house of the wealthy Habert de Montmor. This house became the Montmor Academy in 1657, and its meetings were attended by the secretaries of state, several abbés of the nobility, and other top-ranking officials. The Academy championed the mechanical philosophy and maintained close ties with the Royal Society. In 1666, Louis XIV's minister Colbert reor-ganized the Academy as the French Academy of Sciences. As was the case with the Royal Society, the notion of a value-free science was part of a political and religious campaign to create a stable social and ecclesiastical order throughout Europe. What modern science came to regard as abstract truths, such as the radical separation of matter and spirit, or mind and body, were central to this campaign. The success of the mechanical world view cannot be attributed to any inherent validity it might possess, but (partly) to the powerful political and religious attack on the Hermetic tradition by the reigning European elites.[60]

Just as the Mersenne circle's opposition to Hermeticism took the form of an attack on the occult affiliations of Protestantism, so was the Protestant attack on magic an integral part of its opposition to Catholicism. We have already seen how intimate were the ties between magic and the church on the local level, and how essential these were to the maintenance of its authority. We should not be surprised, then, to discover that the Reformation adopted a deliberately rationalist front. All the sacraments were scrutinized for their magical affiliations. Lists of popes who had allegedly been conjurers were compiled and circulated, and even such practices as saying "God bless you" when a person sneezed were attacked as superstitious claptrap. Ultimately, the attack succeeded. By 1600 the view that God could not be conjured, and that ritual ceremonies (such as transubstantiation) could not have material efficacy, was gaining ground. The idea that physical objects had Mind, or *mana* behind them, and could be altered by exorcism or alchemical procedure, began to be seriously attenuated.[61]

In addition, Protestantism was able to undercut the soteriological claims of Hermeticism with the concept of secular salvation. It is interesting that this concept adopted the structure of magical practice exactly. As we have already noted, the efficacy of the practitioner was seen as being a function of his inner purity or virtue. In the same way, the evidence of grace in, for example, Calvinism, was worldly success. As Weber described at length, money was now viewed as salvation made manifest, the touchstone of real piety. And in the context of nascent capitalism, the concept of personal salvation through internal psychic regeneration, which was now openly advocated by groups such as the Rosicrucians, simply could not compete. For the middle and upper classes, at least, the vacuum left by the Protestant attack on the supernatural could be filled by prayer and worldly success. But since secular salvation was so obviously a "winner's" philosophy, Protestantism was in the position of imposing a doctrine on a populace long used to other types of explanation.[62] Throughout Northern Europe, both the notion of secular salvation and the mechanical philosophy informed the world view of the rising bourgeoisie; it was their spiritual needs alone that would be catered to. The imposition of this new doctrine involved not only oppression of others, but repression of self. The Puritan values of competitiveness, orderliness, and self-control came to typify a world that had previously regarded such behavior as aberrant; or, in the case of Isaac Newton, as frankly pathological.[63] As Christopher Hill puts it, the "preachers knew what they were doing. . . . They were up against 'natural man.' The mode of thought and feeling and repression which they wished to impose was totally unnatural."[64] Today, we have to live with the consequences of their success, and regard it, and the mechanical world view, as "normal." But if Hermeticism does correspond to an archaic substrate in the human psyche, as Jung's work seems to indicate, and if creativity and individuation are drives inherent in human nature, then our modern view of reality was purchased at a fantastic price. For what was ultimately created by the shift from animism to mechanism was not merely a new science, but a new personality to go with it; and Isaac Newton can rightly be seen as a microcosm, or epitome, of these changes. I wish, then, to complete this survey of the collapse of participating consciousness with a separate examination of Newton's life and work in relation to the political and religious events of his day. Only then will we be in a position to assess the cost of the loss of holism in the West and to open the question of what is still possible for those of us who are, both philosophically and psychologically, the heirs of the Newtonian synthesis.

4

The Disenchantment
of the World (2)

For nature is a perpetual circulatory worker, generating fluids out
of solids, and solids out of fluids; fixed things out of volatile, and
volatile out of fixed; subtle out of gross, and gross out of subtle;
some things to ascend, and make the upper terrestrial juices,
rivers, and the atmosphere, and by consequence others to de-
scend for a requital to the former.

—Isaac Newton, from a letter to
Henry Oldenberg, 25 January 1675/6

[I]t seems probable to me that God in the beginning formed matter in solid, massy, hard, impenetrable, movable particles, of
such sizes and figures, and with such other properties and in
such proportion to space as most conduced to the end for which
he formed them. . . . And therefore, that nature may be lasting,
the changes of corporeal things are to be placed only in the various separations and new associations and motions of these permanent particles. . . .

—Isaac Newton, 31st Query to the
Opticks, 4th edition, 1730

Isaac Newton is the symbol of Western science, and the *Principia* may rightly be called the hinge point of modern scientific thought. As we saw in chapter 1, Newton defined the method of science itself, the notions of hypothesis and experiment, and the techniques that were to make rational mastery of the environment a viable intellectual program. Through the public stance adopted by Newton and his disciple Roger Cotes, the positivist conception of truth first advanced by Mersenne was stamped upon the European mind. And although twentieth-century physics has modified the details of the Newtonian synthesis significantly, all modern scientific thinking, if not the character of contemporary rational-empirical thought in general, remains, in essence, profoundly Newtonian.

It was thus with some amazement that, when masses of Newton's manuscripts were auctioned off by his descendants at

Sotheby's in 1936, the British economist John Maynard Keynes read through them and discovered that Newton had been steeped in, if not obsessed by, the occult sciences, particularly alchemy.[1] As a result, Keynes could not avoid making the following judgment:

> Newton was not the first of the age of reason. He was the last of the magicians.... He looked on the whole universe and all that is in it *as a riddle*, as a secret which could be read by applying pure thought to certain evidence, certain mystic clues which God had laid about the world to allow a sort of philosopher's treasure hunt to the esoteric brotherhood. He believed that these clues were to be found partly in the evidence of the heavens and in the constitution of elements (and this is what gives the false suggestion of his being an experimental natural philosopher), but also partly in certain papers and traditions handed down by the brethren in an unbroken chain back to the original cryptic revelation in Babylonia. He regarded the universe as a cryptogram set by the Almighty.[2]

Keynes realized that the eighteenth century had essentially "cleaned Newton up" for public viewing; that the *Principia* and the *Opticks* were but the published portion of a larger quest that had much more in common with the world view of, say, Robert Fludd than with that of a nineteenth-century physicist. But the recent biography of Newton by Frank Manuel, and the brilliant study of Newton and his cultural context by David Kubrin, have shown that to a great extent, Newton cleaned himself up as well.[3] To find the answer to the riddle of gravity on the particulate level, Newton turned to the Hermetic tradition; and he came to see himself, Keynes suggests, as the contemporary representative, indeed even the God-chosen inheritor of that tradition. But for both psychological and political reasons, Newton found it necessary to repress that side of his personality and his philosophy, and to present a sober, positivist face. In significant ways, the evolution of Newton's consciousness reflects not only the fate of the alchemical tradition in Restoration England, but also the evolution of Western consciousness in general. Indeed, Manuel has suggested that his personality and outlook were but extreme expressions of the age.[4]

Newton's childhood was characterized by an intense dose of the separation anxiety that is a part of all of our early lives and that later serves as a model for the sensation of bodily responses that occur whenever we face object-loss. Newton's father died three months before he was born, and his mother remarried when he was just about three years of age. She went to live a mile and a half away with her new husband, the Reverend Barnabas Smith, leaving Isaac with his grandmother in Woolsthorpe, Lincolnshire, the town of his birth. She returned to Woolsthorpe only when her second husband died, by which time Newton was about eleven. Hence Newton was quite literally abandoned during a crucially formative period, after a period in which his mother had been the sole parent. As a result, writes Manuel,

> his fixation upon her was absolute. The trauma of her original departure, the denial of her love, generated anguish, aggressiveness, and fear. After the total possession—undisturbed by a rival, not even a father, almost as if there had been a virgin birth—she was removed and he was abandoned.

"The loss of his mother to another man," continues Manuel, "was a traumatic event in Newton's life from which he never recovered." Newton recorded in one of his adolescent notebooks "sins" such as "threat[e]ning my father and mother Smith to burne them and the house over them," and "wishing death and hoping it to some."

It should also be noted that Newton's belief that he was part of the *aurea catena*, the "golden chain" of magi, or unique figures designated by God in each age to receive the ancient Hermetic wisdom, was reinforced by the circumstances of his birth. He was born prematurely, on Christmas Day 1642, and was not expected to live. Indeed, that particular parish had a high rate of infant mortality, and Newton later believed that his survival (as well as his escaping the ravages of the plague while still a young man) signified divine intervention. The same parish, according to Manuel, also credited some form of the widespread belief that a male child born after his father's death is endowed with extraordinary powers. This attitude, combined with Newton's great fear of object-loss, produced his peculiar stance with respect to past and present thinkers. Moses, Thoth, Thales, Hermes, Pythagoras, and others like them enjoyed his praise; contemporary scientists were by and large a threat. Newton went into extreme rages in his arguments over priority with men such as Hooke and Leibniz, and regarded the system of the world described in the *Principia* as his personal property. He was certain that "God revealed himself to only one prophet in each generation, and this made parallel discoveries improbable." At the bottom of one alchemical notebook Newton inscribed as an anagram of his Latin name, Isaacus Neuutonus, the phrase: *Jeova sanctus unus*—Jehovah the holy one.

Along with these psychological traits, Newton manifested those common to Puritan morality: austerity, discipline, and above all, guilt and shame. "He had a built-in censor," says Manuel, "and lived ever under the Taskmaster's eye. . . ." Such conclusions emerge from a study of Newton's adolescent exercise notebooks, which include sentences chosen for translation into Latin in the manner of free association—sentences in which dread, self-disparagement, and loneliness abound as themes. Hence:

> A little fellow.
> He is paile.
> There is noe roome for mee to sit.
> What imployment is he fit for?
> What is hee good for?
> He is broken.
> The ship sinketh.
> There is a thing which trobeleth mee.
> He should have been punished.
> No man understands mee.
> What will become of me.
> I will make an end.
> I cannot but weepe.
> I know not what to doe.

These are remarkable sentences for a youth to choose for Latin exercises, indeed, the selection is almost unbelievable. "In all these youthful scribblings," writes Manuel,

> there is an astonishing absence of positive feeling. The word *love* never appears, and expressions of gladness and desire are rare. . . . Almost all the statements are negations, admonitions, prohibitions. The climate of life is hostile and punitive.

Had history heard nothing more from Isaac Newton, these notebook entries would amount to nothing more than a psychiatric curiosity. But we are talking about the creator of the modern scientific outlook, and that outlook, the insistence that everything be totally predictable and rationally calculable ("kill anything that moves," as Philip Slater puts it) cannot be separated from its pathological basis. "A chief source of Newton's desire to know," writes Manuel, "was his anxiety before and his fear of the unknown." "Knowledge that could be mathematicized ended his quandaries... [The fact] that the world obeyed mathematical law was his security."

> To force everything in the heavens and on earth into one rigid, tight frame from which the most minuscule detail would not be allowed to escape free and random was an underlying need of this anxiety-ridden man. And with rare exceptions, his fantasy wish was fulfilled during the course of his lifetime. The system was complete in both its physical and historical dimensions. A structuring of the world in so absolutist a manner that every event, the closest and the most remote, fits neatly into an imaginary system has been called a symptom of illness, especially when others refuse to join in the grand obsessive design. It was Newton's fortune that a large portion of his total system was acceptable to European society as a perfect representation of reality, and his name was attached to the age.[5]

The schizophrenic, wrote the anthropologist Géza Róheim, is the magician who has failed.[6] Despite his eventual nervous breakdown, Newton was no psychotic; but that he bordered on a type of madness, and allayed it with a totally death-oriented view of nature, is beyond doubt. What is significant, however, is not his view of nature itself, but the broad agreement that it found, the excitement that it generated. Newton was the magician who succeeded. Instead of remaining some sort of isolated crank, he was able to get all of Europe "to join in the grand obsessive design," becoming president of the Royal Society and being buried, in 1727, amidst pomp and glory in Westminster Abbey in what was literally an international event. With the acceptance of the Newtonian world view, it might be argued, Europe went collectively out of its mind.

Where does Newton's Hermeticism fit into all of this? We have already seen that he regarded himself as the inheritor of an archaic tradition, what D. P. Walker has called the *prisca theologia* (ancient theology), a collection of church-related texts believed, during the Renaissance, to have been inspired by knowledge that dated back to the time of Moses and which embodied the secrets of matter and the universe.[7] Newton's alchemical library was indeed large, and his alchemical experiments were a major feature of his life down to 1696 when he moved to London to become master of the Mint. Newton was connected to alchemy by something that was inte-

121

grally related to his megalomania about inheriting the sacred tradition: his conviction that matter was not inert but required an active, or hylarchic, principle for its motion. In alchemy Newton hoped to find the microcosmic correlate to gravitational attraction, which he had already established on the macrocosmic level. As Gregory Bateson has rightly remarked, Newton did not discover gravity; he *invented* it.[8] This invention, however, was part of a much larger quest: Newton's search for the system of the world, the secret of the universe—an ancient riddle stretching back, as Keynes said, to the Babylonians. The Hermetic tradition was thus the framework of early Newtonian thought, and gravity merely a name for the hylarchic principle that he was certain had to exist.[9] Newton was first and foremost the alchemist Keynes saw in him, then. Over the years, however, as the result of a self-repression that had an important political motivation behind it, he gradually evolved into a mechanical philosopher.

English interest in alchemy, and mysticism in general, became intense during the period of Newton's childhood, the Civil War and after. More alchemical and astrological texts were translated into English during 1650–60 than in the entire preceding century.[10] The reasons for this increased interest were largely political. Even today, one's view of matter and force is inevitably a religious question; and in the context of the seventeenth century, religious questions were typically political issues as well. At one level, the Civil War signified the breakdown of a feudal economy; the opposition of the new bourgeoisie, with its laissez-faire outlook, to the monopolistic practices of the crown. This economic struggle was reflected politically in the conflict between Royalists and Parliamentarians, and religiously in the triumph of Puritanism. But the war had another dimension, now recovered in the work of Christopher Hill: the attempt, on the part of a vast number of sects, to fight the crown, and later the Parliamentarians, with the ideology of communism, or what Engels called utopian socialism, and to argue for direct knowledge of God as opposed to salvation either through works or blind faith.[11] The religion of these numerous groups—Levellers, Diggers, Muggletonians, Familists, Behmenists, Fifth Monarchy Men, Ranters, Seekers—was in many cases some combination of Hermeticism, Paracelsism, or soteriological alchemy, and hence they were often linked in the public mind with what was called "enthusiasm," that is, immoderation in religious beliefs, including possession by God or prophetic frenzy. All mystical experiences, naturally enough, came under this heading, and many of the radicals had clearly had such ecstatic insights.[12] It "was among the mystical sects," writes Keith Thomas, "that alchemy struck some of its deepest roots."[13] While there have been no studies demonstrating the actual extent of such beliefs among the lower classes and radical groups, there is little problem in demonstrating that such an association was made in the public (especially middle-class) mind of the time. At the center of these beliefs was a view of nature directly opposed to the new science: the notion that God was present in everything, that matter was alive (pantheism); that change occurred via internal conflict (dialectical reason) rather than rearrangement of parts; and that— in contrast to the hierarchical views of the Church of England—any individual could attain enlightenment and have direct experience of the Godhead (soteriological alchemy). The attempt of the lower classes to hang onto Hermetic notions reflected the class split described by Keith Thomas, who observed that the Protestant/ rationalist attack on magic left the middle class with secular salvation, and the lower classes (in a context of enclosures and accelerating poverty) with nothing. During this period, then, Hermeticism had an unmistakably socialist edge.[14]

The political threat inherent in the occult world view, however, went far beyond the attack on property and privilege espoused by most of these radical sects. It included: outright atheism; rejection of monogamy and an affirmation of the pleasures of the body; demands for religious toleration, as well as for the abolition of the tithe and the state church; contempt for the regular clergy; and rejection of any notions of hierarchy, as well as of the concept of sin. The ties between occult and revolutionary thought can be seen in a whole spectrum of leading radicals, but, as already noted, the popular impression that communism, libertinism, heresy, and Hermeticism were part of some vast conspiracy is amply documented in the numerous statements made on the subject by clergymen.[15] This intense political/occult ferment, and the fear of it, received full expression in the 1640s. In the 1650s, however, the tide began to turn; and after the Restoration, the mechanical philosophy was seen by the ruling elites as the sober antidote to the enthusiasm of the last two decades. From 1655 onward there was a series of conversions to the mechanical philosophy by men who had previously been sympathetic to alchemy.

These conversions were thus part of the reaction against enthusiasm on the part of the propertied classes and leading members of the Church of England, groups that coalesced in the Royal Society itself. Thomas Sprat, in the earliest history of the Society (1667), viewed the mechanical philosophy as helping to instill respect for law and order, and claimed that it was the job of science and the Royal Society to oppose enthusiasm. Men like Charleton and Boyle, key figures in the conversion to mechanism, worried about the influence of an alchemist like Jacob Boehme among English radicals. They feared that the proliferation of religions based on mystical insight or individual conscience would end in no religion at all. "Elevation of the mechanical philosophy above the dialectical science of radical 'enthusiasts,'" writes Christopher Hill, "reciprocally helped to undermine such beliefs."[16]

As the reader might imagine, Newton, who had his most brilliant insights regarding the system of the world in 1666, was in something of a quandary. It must have been as evident to him as to any student at Restoration Cambridge, writes Kubrin, "that Hermetic knowledge was widely viewed by his contemporaries as an inducement to enthusiasm, and that extreme caution should be exercised with regard to such ideas." At the same time, he saw himself as the inheritor of the sacred tradition, and was convinced that the answer to the riddle of the system of the world was buried within it. What Newton did, then, was to delve deeply into the Hermetic wisdom for his answers, while clothing them in the idiom of the mechanical philosophy.

The centerpiece of the Newtonian system, gravitational attraction, was in fact the Hermetic principle of sympathetic forces, which Newton saw as a creative principle, a source of divine energy in the universe. Although he presented this idea in mechanical terms, his *unpublished* writings reveal his commitment to the cornerstone of all occult systems: the notion that mind exists in matter and can control it (original participation). In his letter to Henry Oldenburg, secretary of the Royal Society, cited in the epigraph to this chapter, Newton states that "nature is a perpetual circulatory worker," and then offers a description of nature's mode of operation—separating the gross from the subtle, the volatile from the fixed, and so on—which is alchemy pure and simple. Draft versions of published work contain statements that were not publicly heard in the West, in the modern period, until Lamarck and Blake: "all matter duly formed is attended with signes of life"; "nature delights in transformations"; the world is "God's sensorium," and so on. His writings abound with alchemical notions, such as fermentation and putrefaction, or the "sociableness" and "unsociableness" of various substances for each other; and some of these statements even made their way into the famous 31st Query of the *Opticks*. [17] As R. S. Westfall puts it, alchemy was Newton's most enduring passion, and the *Principia* something of an interruption of this larger quest. [18]

Even some of Newton's published work (like the 31st Query) reveals his intense interest in the occult. The reader may be surprised to learn that Newton wrote on the ancient temple of King Solomon, and speculated on the size of that ancient measure, the cubit. [19] The notion that the secrets of the universe were contained in the mathematical relationships built into the structure of ancient holy buildings was a part of the Hermetic tradition, one that is making something of a comeback with the current vogue of "pyramid power." Indeed, Newton had a similar interest in the Great Pyramid of Cheops, and as with his attempt to use alchemical experiments to validate the theory of gravity, this interest was much more than an unrelated hobby. Newton was later to state that Egyptian priests knew the very secrets of the cosmos which he had revealed in the *Principia*.

Newton's retreat from these views, as Kubrin is able to show, occurred in the context of a revival of Hermetic ideas in the late 1670s and the 1680s, the years leading up to the Glorious Revolution. [20] Leveller and republican sentiments emerged once again, and a leading proponent of the new Hermeticism, especially in the 1690s, was one John Toland, who had studied with the Newtonian scholar David Gregory. Toland saw the animistic notions lurking in Newton's work and pointed to them in his own publications, claiming that nature was transformative and infinitely fecund, and drawing an analogy to the political arena. Newton's dilemma was that he secretly agreed with Toland's theory of matter and force, and had in fact held these views for decades. It thus became imperative for him to dissociate himself from these ideas; but this necessarily meant changing his mind about them in what amounted to a rigorous self-censorship. His disciple Samuel Clarke was entrusted with the job of attacking Toland in a set of sermons published in 1704, and when Clarke translated the *Opticks* into Latin two years later, the phrase, the world is "God's sensorium," was altered to read, is "like God's sensorium." [21] Statements such as "we cannot say that all Nature is not alive" were withdrawn before publications went to press; and most importantly, Newton adopted the position that matter was inert, that it changed not dialectically (i.e., internally) but through rearrangement alone. Thus in the quotation from the *Opticks* cited at the beginning of this chapter, Newton gives as his purpose "that nature may be lasting"; in other words, that it may be stable, predictable, regular—like the social order ought to be. As a young man, Newton had been fascinated by the fecundity of nature. Now, its alleged rigidity was somehow all-important.

In the modern empirical sense, there was nothing "scientific" about this shift from Hermeticism to mechanism. The change was not the result of a series of careful experiments on the nature of matter, and indeed, it is no more difficult to visualize as a living organism than it is to see it as a dead, mechanical object.[22] And at the risk of stretching a point somewhat, it seems to me, following Kubrin's argument, that two things must be noted about this transformation, in addition to its nonscientific character. First, the forces that triumphed in the second half of the seventeenth century were those of bourgeois ideology and laissez-faire capitalism. Not only was the idea of living matter heresy to such groups; it was also economically inconvenient. A dead earth ruptures the delicate ecological balance that was maintained in the alchemical tradition, but if nature is dead, there are no restraints on exploiting it for profit. Loving cultivation becomes rape; and that, to me, is most clearly what industrial society in general (not just capitalism) represents. That the current breakdown of such societies, at least in the West, is being accompanied by an occult revival, with all its good and bad aspects, is hardly surprising.

Second, the triumph of the Puritan view of life, which concomitantly repressed sexual energy and sublimated it into brutalizing labor,[23] helped to create the "modal personality" of our time—a personality that is docile and subdued in the face of authority, but fiercely aggressive toward competitors and subordinates. The severely repressed Newton, as Blake pointed out, was everyman; and various paintings of Newton done over the period 1689 to 1726 (Plates 12–15) reveal an increasing amount of what Wilhelm Reich brilliantly termed "character armor." In the earliest painting, the "Hermetic" Newton retains (despite his childhood) a gentle, ethereal quality that the artist has captured quite beautifully. In the end, however, we see the rigidity of the mechanical world view, the Newton who denied his own internal principles—what Rilke

Plate 12. Isaac Newton, 1689. Portrait by Godfrey Kneller. Lord Portsmouth and the Trustees of the Portsmouth Estate.

Plate 13. Isaac Newton, 1702. Portrait by Godfrey Kneller. Courtesy, National Portrait Gallery, London.

127

Plate 14. Isaac Newton, ca. 1710. Portrait by James Thornhill. By permission of the Master and Fellows of Trinity College, Cambridge.

Plate 15. Isaac Newton, 1726, the year before his death. Mezzotint by John Faber, after painting by John Vanderbank. Courtesy, Prints Division, The New York Public Library, Astor, Lenox and Tilden Foundations.

128

called the "unlived lines in our bodies"[24]—for the sake of social approval and outward conformity. We see, in effect, the tragedy of modern man.[25]

Finally, as a number of writers have pointed out, just as the lower classes were suppressed at the level of work and labor, so did the middle and upper classes keep themselves in check at the level of literary and intellectual activity. The attack on enthusiasm was breathtakingly successful, and is reflected in the poetry of the eighteenth century (the carefully contrived couplets of Dryden and Pope) as well as the notion of classical scholarship itself. "The classics!" cried Blake. "It is the classics, and not Goths nor monks, that desolate Europe with wars."[26] In his painting of Newton, carving up the world with a compass (Plate 16), Blake tried to show

Plate 16. William Blake, *Newton* (1795). The Tate Gallery, London.

129

the blindness of this orientation to nature; and nowhere did he say it better than in his verse letter to Thomas Butts (1802):

> Now I a fourfold vision see,
> And a fourfold vision is given to me;
> 'Tis fourfold in my supreme delight
> And threefold in soft Beulah's night
> And twofold always. May God us keep
> From Single vision & Newton's sleep![27]

Newton is pictured in Blake's painting sitting at the bottom of the Sea of Space and Time. The polyp near his left foot symbolizes, in Blake's mythology, "the cancer of state religion and power politics," while Newton stares at his diagram "with the catatonic fixity of 'single vision'...."[28]

Blake's attack on the Newtonian world view raises a question that Hill has made the theme of *The World Turned Upside Down:* how can we be so sure that the way things are is right side up? Bourgeois society, he notes, was a powerful civilization, producing great intellects in the Newtonian and Lockean mold. But, he adds, it was

> the world in which poets went mad, in which Locke was afraid of music and poetry, and Newton had secret, irrational thoughts which he dared not publish....
>
> Blake may have been right to see Locke and Newton as symbols of repression. Sir Isaac's twisted, buttoned-up personality may help us to grasp what was wrong with the society which deified him.... This society, which on the surface appeared so rational, so relaxed, might perhaps have been healthier if it had not been so tidy, if it had not pushed all its contradictions underground: out of sight, out of conscious mind.... What went on underground we can only guess. A few poets had romantic ideas out of tune with their world; but no one needed to take them too seriously. Self-censored meant self-verifying.[29]

"Great though the achievements of the mechanical philosophy were," Hill writes at another point, "a dialectical element in scientific thinking, a recognition of the 'irrational' (in the sense of the mechanically inexplicable) was lost when it triumphed, and is having to be painfully recovered in our own century."[30]

The emphasis here is on the word "painfully." In Chapter 3, I discussed the role of surrealist art in attempting to liberate the unconscious. But because the unconscious is so repressed, its great mouthpiece in postwar Europe and America has become not art, but madness. Without going into too much detail, it is necessary to point out that a major part of the psychotic experience is the return to the perception of the world in Hermetic terms. That madness is the best route to this perception I tend to doubt; but the fact that madness triggers the premodern epistemology of resemblance does suggest that the insane are onto something we have forgotten, and that (cf. Nietzsche, Laing, Novalis, Hölderlin, Reich . . .) our sanity is nothing but a collective madness.

Although it would take extensive clinical studies of insanity to establish this argument, even a casual review of the case histories described by Laing in *The Divided Self* tends to substantiate it.[31] In general, says Laing, having a disembodied self creates a sense of merger or confusion at the interface between inside and outside. As in soteriological alchemy or mystical experience, the subject/ object distinction blurs; the body is not felt as being separate from other things or people. One of Laing's patients, for example, did not distinguish between rain on her cheek and tears. She also worried that she was destructive, in the sense that if she touched anything, she would literally damage it (antipathy theory). Schizophrenics occasionally demonstrate a belief that inanimate objects contain extraordinary powers, and Laing describes the case of a man who, while on a picnic, undressed and walked into a nearby river, declaring that he had never loved his wife and children, pouring water on himself repeatedly, and refusing to leave the river until he had been "cleansed." Here we have the original notion of baptism, the belief that water bears the impressed virtue of God (doctrine of signatures), and thus has healing powers. Another patient practiced various techniques to "recapture reality," such as repeating phrases she regarded as real over and over in the hope that their "realness" would rub off on her (sympathy theory, notion of *mana*). Finally, as I indicated in Chapter 3, Laing's own method is alchemical in that it follows the notion of participating consciousness, or sympathy theory. All humanistic therapies, in fact, are rooted in original participation. The use of art, dance, psychodrama, meditation, body work, and the like ultimately boils down to a merger of subject and object, a return to poetic imagination or sensuous identification with the environment. In the last analysis, the good therapist is nothing more than the master alchemist to his or her patients, and effective therapy is essentially a return to the inherent, organic order that magic represented. The classification schemes of modern science, their Linnaean order and precision, purport to arise from the ego alone, to be fully rational-empirical. They thus represent a logical order that is *imposed* on nature and the human psyche. As a result, they violate something that magic, for all its technological limitations, had the instinctive wisdom to preserve.

Madness is, in the end, a statement about logical categories, and its reversion to the structure of premodern thought represents a revolt against the reality principle that it sees as crushing the human spirit. The increasing incidence of madness in our time reflects the desperate need for the recovery of dialectical reason. Does alchemy, or technology, represent the altered state of consciousness? Is material production, or human self-realization, most consonant with true human needs? Is subjugation of the earth, or harmony with it, the best way to proceed? I would submit that there is only one answer to these questions, and only one conclusion to our survey of the disenchantment of the world: in the seventeenth century, we threw out the baby with the bathwater. We discounted a whole landscape of inner reality because it did not fit in with the program of industrial or mercantile exploitation and the directives of organized religion. Today, the spiritual vacuum that results from our loss of dialectical reason is being filled by all kinds of dubious mystical and occult movements, a dangerous trend that has actually been encouraged by the ideal of the disembodied intellect and the classical scholarship that Blake rightly found revolting. Modern science and technology are based not only on a hostile attitude toward the environment, but on the repression of the body and the unconscious; and unless these can be recovered, unless participating consciousness can be restored in a way that is scientifically (or at least rationally) credible and not merely a relapse into naive animism, then what it means to be a human being will forever be lost.

The remainder of this book will be devoted to an exploration of such options.

5

Prolegomena to Any Future Metaphysics[1]

Perhaps we need to be much more radical in the explanatory hypotheses considered than we have allowed ourselves to be heretofore. Possibly the world of external facts is much more fertile and plastic than we have ventured to suppose; it may be that all these cosmologies and many more analyses and classifications are genuine ways of arranging what nature offers to our understanding, and that the main condition determining our selection between them is something in us rather than something in the external world.

—E. A. Burtt, *The Metaphysical Foundations of Modern Science*

In previous chapters we have discussed the modern scientific outlook, demonstrated its relationship to certain social and economic developments, and examined the psychological landscape that it destroyed. This analysis suggests that the Western world paid a high price for the triumphs of the Cartesian paradigm and that there are severe limits to it in terms of human desirability. Indeed, even its objective accuracy can be debated for, as we have seen, its triumph over the metaphysics of participating consciousness was not a scientific but a political process; participating consciousness was rejected, not refuted. As a result, we are forced to consider the possibility that modern science may not be epistemologically superior to the occult world view, and that a metaphysics of participation may actually be more accurate than the metaphysics of Cartesianism. A number of scientific thinkers, including Alfred North Whitehead, have argued this thesis in one form or another and, as early as 1923, the psychologist Sándor Ferenczi called for the "re-establishment of an animism no longer anthropomorphic."[2] Yet our culture hangs on to mechanism, and to all of the problems and errors it involves, because there is no returning to Hermeticism and—apparently—no going on to something else.

I have promised to devote the second half of this book to "something else," and in subsequent chapters I shall enlarge on what might serve as a post-Cartesian world view. Before contemplating an alternative, however, it is necessary to elaborate on a key weakness in the epistemology of modern science—the fact that it contains participating consciousness even while denying it. It is this denial that has created the characteristic paradoxes of scientific thought, notably its radical relativism, and which has also made it impossible for orthodox scientific thinking to evolve in new directions, such as those suggested by quantum mechanics. I maintain that an understanding of the stubborn persistence of participating consciousness can help us to solve the problem of radical relativism and also suggest some theoretical underpinnings for a post-Cartesian science. The arguments I am going to advance, then, are as follows:

(1) Although the denial of participation lies at the heart of modern science, the Cartesian paradigm as followed in actual practice is riddled with participating consciousness.

(2) The deliberate inclusion of participation in our present epistemology would create a new epistemology, the outlines of which are just now becoming visible.

(3) The problem of radical relativism disappears once participation is acknowledged as a component of all perception, cognition, and knowledge of the world.

Fortunately for this discussion, point (1) is the central focus of two recent and brilliant critiques of modern science: Michael Polanyi's *Personal Knowledge,* and Owen Barfield's *Saving the Appearances.*[3] Polanyi's major thesis is that in attributing truth to any methodology we make a nonrational commitment; in effect, we perform an act of faith. He demonstrates that the coherence possessed by any thought system is not a criterion of truth, but "only a criterion of *stability.* It may [he continues] equally stabilize an erroneous or a true view of the universe. The attribution of truth to any particular stable alternative is a fiduciary act which cannot be analysed in non-commital terms."[4] The faith involved, according to Polanyi, arises from a network of unconscious bits of information taken in from the environment which form the basis of what he calls "tacit knowing." What exactly does this concept mean?

We already have alluded to the notion of a gestalt perception of reality, of finding in nature what you seek. Philosopher Norwood Russell Hanson used the illustrations given in Figures 10 and 11 to make this point:[5]

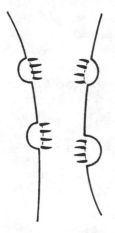

Figure 10. N. R. Hanson's illustration of gestalt perception: tree with burls vs. bear climbing up trunk (from Hanson, *Patterns of Discovery*, p. 12)

Figure 11. N. R. Hanson's illustration of gestalt perception: flock of birds vs. herd of antelope (from Hanson, *Patterns of Discovery*, p. 13)

In Figure 10, do you see a bear climbing up the other side of a tree, or a tree trunk with burls on it? Do you see a flock of birds in Figure 11, or a herd of antelope? Would people who had never seen antelope, but only birds, be able to regard Figure 11 as a picture of antelope? Polanyi's general point is that at a very early age we learn, or are trained, to put reality together in certain ways ("figurate" it, in Barfield's terminology), and that the indoctrination is not merely cultural but also biological. Thus on a conscious level we largely spend our lives finding out what we already know on an unconscious level. Alternative realities are screened out by a process that the American psychiatrist Harry Stack Sullivan used to call "selective inattention," and which has since been relabeled "cognitive dissonance." Thus "antelope" people would presumably find "bird" people incomprehensible. Any articulated world view, in fact, is the result of unconscious factors that have been culturally filtered and influenced, and is thus to some extent radically disparate from any other world view.

The question that concerns us here is how we are trained into a mode of seeing. Polanyi points out that the scientist learns his craft in the same way a child learns a language. Children are born polyglots: they naturally have German gutterals, French nasals, Russian palatals, and Chinese tonals. They cannot remain this way for long, however, for to learn a particular language is simultaneously to unlearn the sounds not common to that language. English, for example, does not have the Russian palatal sound, and the English-speaking child ultimately loses the ability to pronounce words in a genuinely Russian manner. The awareness here is subsidiary, or even subliminal. As in bicycle riding, so in speaking, we learn to do something without actually analyzing or realizing what it is we are learning. Science has similarly an ineffable basis; it too is picked up by osmosis.[6]

Polanyi's best example of this process, taken perhaps from his own experience, is that of the study of X-ray pathology, and is worth quoting in full.

Think [he writes] of a medical student attending a course in the X-ray diagnosis of pulmonary diseases. He watches in a darkened room shadowy traces on a fluorescent screen placed against a patient's chest, and hears the radiologist commenting to his assistants, in technical language, on the significant features of these shadows. At first the student is completely puzzled. For he can see in the X-ray picture of a chest only the shadows of the heart and ribs, with a few spidery blotches between them. The experts seem to be romancing about figments of their imagination; he can see nothing that they are talking about. Then as he goes on listening for a few weeks, looking carefully at ever new pictures of different cases, a tentative understanding will dawn on him; he will gradually forget about the ribs and begin to see the lungs. And eventually, if he perseveres intelligently, a rich panorama of significant details will be revealed to him: of physiological variations and pathological changes, of scars, of chronic infections and signs of acute disease. He has entered a new world. He still sees only a fraction of what the experts can see, but the pictures are definitely making sense now and so do most of the comments made on them. He is about to grasp what he is being taught; it has clicked.[7]

"He has entered a new world." Polanyi describes a process that is not really rational but existential, a groping in the dark after the fall through Alice's rabbit hole has occurred. There is no *logic* of scientific discovery here, but rather an act of faith that the process will lead to learning, and on the basis of the student's commitment, it does.

It is also important to note, in this example, that the actual learning process violates the Platonic/Western model of knowledge, which insists that knowledge is obtained in the act of distancing oneself from the experience. Our hypothetical medical student knew absolutely nothing when he stood outside the procedures. Only with his submergence in the experience did the photographs begin to take on any meaning at all. As he forgot about himself, as the independent "knower" dissolved into the X-ray blotches, he found that they began to appear meaningful. The crux of such learning is the Greek concept of *mimesis*, of visceral/poetic/erotic identification. Even from Polanyi's verbal description, we can almost touch the willowy blotches on the warm negative, smell the photographic developer in the nearby darkroom. This knowledge was clearly participated.

Rationality, as it turns out, begins to play a role only *after* the knowledge has been obtained viscerally. Once the terrain is familiar, we reflect on how we got the facts and establish the methodological categories. But these categories emerge from a tacit network, a process of gradual comprehension so basic that they are not recognized as "categories." As Marshall McLuhan once remarked, water is the last thing a fish would identify as part of its environment, if it could talk. In fact, the categories start to blur with the learning process itself; they become "Reality," and the fact that the existence of other realities may be as possible as the existence of other languages usually escapes our notice. The reality system of any society is thus generated by an unconscious biological and social process in which the learners in that society are immersed. These circumstances, says Polanyi, demonstrate "the pervasive participation of the knowing person in the act of knowing by virtue of an art which is essentially inarticulate." I can speak of this knowledge, but I cannot do so adequately.[8]

For Polanyi, then, a phrase such as "impersonal" or "objective knowledge" is a contradiction in terms. He argues that all knowing takes place in terms of meaning, and thus that the knower is implicated in the known. To this I would add that what constitutes knowledge is therefore merely the findings of an agreed-upon methodology, and the facts that science finds are merely that—

facts that *science* finds; they possess no meaning in and of them-selves. Science is generated from the tacit knowing and subsidiary awareness peculiar to Western culture, and it proceeds to construct the world in those particular terms. It if is true that we create our reality, it is nevertheless a creation that proceeds in accordance with very definite rules—rules that are largely hidden from con-scious view.

Participating consciousness is even more pervasive than Polanyi's example of the X-ray student would suggest. To see this, let us follow Barfield and define *figuration* as representation, that is, the act by which we transform sensations into mental pictures.[9] The process of thinking about these "things," these images, and their relationships with each other (a process commonly called conceptualization) can be defined as *alpha-thinking*. In the process of learning, figuration gradually becomes alpha-thinking; in other words, our concepts are really habits. Our X-ray student at first formed mental pictures of the blotches or shadows on the screen, then learned to identify cancer and tuberculosis. His instructors, however, immediately and unthinkingly saw cancer and TB with-out experiencing the blotches in the same way he did. Similarly, when I hear a bird singing, I form some sort of mental image of the sound and try to sort it out. My friend, a professional ornithologist, goes through no such process. He hardly even hears the notes. What comes to his mind, quite automatically, is "thrush." Thus, at least in his professional capacity, he is doing alpha-thinking all the time. He is beyond figuration, whereas I am still struggling with it. It would be more correct to say that he figurates in terms of con-cepts rather than sensations and primary data. He does, then, participate the world (or at least the bird world), but for the most part as a collection of abstractions.

Now the crux of the matter is this: in terms of the dominant reality system, we are all ornithologists. We experience an agreed-upon set of alpha-thoughts, or what Talcott Parsons calls "glosses," instead of the actual events. In short, we continue the process of figuration which began in the learning stages, but it becomes automatic and conceptual rather than dynamic and con-crete.

Peter Achinstein provides a good example of this phenomenon in his book *Concepts of Science*. Let us say that you and I are sitting on the steps of an old farmhouse in the country one summer night, looking down the dusty road that leads to the house. As we sit there, we see a pair of headlights coming up the road. Having nothing more profound on my mind at the moment, I turn to you and say, "There's a car coming up the road." You are silent for a moment and then ask me, "How do you know it's a car? After all, it could be two motorcycles riding side by side." I reflect on this, and then decide to modify my original statement. "You're right. Either there's a car coming up the road, or two motorcycles riding side by side at the same speed." "Hold on," you reply. "That's not necessarily the case either. It could be two large bunches of fireflies." At this point, I may wish to draw the line. We could, after all, do this all night. The point is that in our culture, two parallel lights moving at the same speed along a road at night invariably denote an automobile. We do not really experience (figurate) the lights in any detail; instead we figurate the concept "car." Only an infant (or a poet, or a painter) might figurate the experience in the rich possibility of its detail; only a student figurates X-ray images.[10] Every culture, every subculture (ornithology, X-ray pathology) has a network of such alpha-thoughts, because if we had to figurate everything, we would never be able to construct a science, nor any model of reality. But such a network is a *model*, and we tend to forget that. In Alfred Korzybski's famous dictum (*Science and Sanity*, 1933), "the map is not the territory." After all, what if the lights *were* fireflies?[11]

This confusion of map with territory is what we have called nonparticipating consciousness. Alpha-thinking necessarily involves the absence of participation, for when we think about anything (except in the initial stages of learning) we are aware of our detachment from the thing thought about. "The history of alpha-thinking," writes Barfield, "accordingly includes the history of science, as the term has hitherto been understood, and reaches its culmination in a system of thought which only interests itself in phenomena to the extent that they can be grasped as independent of consciousness."

As we saw in Chapter 3, this distancing of mind from the object of perception was precisely the historical project of the Jews and the Greeks. The Scientific Revolution was the final step in the process, and henceforth all representations in the Western reality system became what Barfield calls "mechanomorphic." Construing reality mechanically is, however, a way of participating the world, but it is a very strange way, because our reality system officially denies that participation exists. What then happens? It ceases to be conscious because we no longer attend to it, writes Barfield, but it does not cease to exist. It does, however, cease to be what we have called *original* participation. Making an abstraction out of nature is a particular way of participating it. Just as the ex-lovers who refuse to have anything to do with one another really have a powerful type of relationship, so the insistence that subject and object are radically disparate is merely another way of relating the two. The problem, the strangeness, lies in the denial of participation's role, not only because the learning process itself necessarily involves *mimesis*, but because as long as there is a human mind, there will be tacit knowing and subliminal awareness.

It might be argued that African tribesmen (for example) are involved in alpha-thinking as much as we are. Once past his apprenticeship, the witch doctor spends much of his time identifying the various members of the spirit world according to a formula. Despite this, the "primitive" slides quite naturally between figuration and alpha-thinking, or in our terminology, between the unconscious and the conscious mind; and he probably spends most of his time experiencing, rather than abstracting. Even if he wished to shut the unconscious out, it would not be possible, because for him the spirits are real and (despite any ritualized system) frequently experienced on a visceral level. His experience of nature constantly creates joy, anxiety, or some intermediate bodily reaction; it is never a strictly cerebral process. He may often be frightened by his environment or by things in it, but he is never alienated by it. There are no Sartres or Kafkas in such cultures any more than there were in medieval Europe. The "primitive" is thus in touch with what Kant called the *Ding an sich*, the thing in itself, in the same way as was the denizen of ancient Greece or (to a lesser extent) medieval Europe. We, on the other hand, by denying both the existence of spirits and the role of our own spirit in our figuration of reality, are out of touch with it. Yet it is the case, as Barfield notes, that in *any* culture "the phenomenal world arises from the relationship between a conscious and an unconscious and that evolution is the story of the changes that relationship has undergone and is undergoing." Denying that the unconscious plays a role in our conceptualization of reality may be a strange way of relating to it, but it is still a way of relating, and it does not erase tacit knowing. Modern textbooks still project the image of a formally applied "scientific method," a method in which any notion of participating consciousness would be tantamount to heresy. Yet the disparity between official image and actual practice is enormous; and as science has perhaps dimly realized, the excommunication of the heresy would bring down the rest of the church in its wake.

The dimensions of this paradox are thrown into sharp relief when we reflect on the unexpected resurfacing or participating consciousness in modern physics in the 1920s. I am referring to the emergence of quantum mechanics, whose theoretical basis involves a full-scale break with the epistemology of Western science. Since the appearance of quantum mechanics is analogous to Ptolemaic astronomy suddenly finding Copernicus in its camp, we should not be surprised that the scientific establishment has managed to ignore the embarrassing intruder for more than five decades. There is, nevertheless, a voluminous literature on the subject, much too extensive to discuss at length here. Instead, I wish to summarize briefly the philosophical implications that can and have been drawn from this branch of physics.[12]

Two concepts are absolutely essential to the epistemology of classical (including Einsteinian) physics. The first is the notion that all reality is ultimately describable in terms of matter and motion; that the position of material particles, and their momentum (mass times velocity), is the basic reality of the phenomenal world. The second point is that ours is a nonparticipating consciousness: the phenomena of the world remain the same whether or not we are present to observe them; our minds in no way alter that bedrock reality. The first of these concepts is the basis of strict causality, or determinism, and it was perhaps best expressed by the French mathematician Pierre Simon de Laplace in 1812. Our physics is such, he said, that if it were possible to know the position and momentum of all the particles in the universe at any one time, we could then calculate their position and momentum at any other time, past or future. The second concept, the conviction that the experimenter is not part of his experiment, affirms the materialism of the first point, and also guarantees that experiments are formally replicable. If, for example, a scientist claimed that by simply concentrating on cubes (e.g., dice) that have been mechanically dropped down a chute, he could influence their spatial pattern, and if his claim turned out to be valid, he would not only have disproved the content of this aspect of physics, he would have destroyed the theoretical basis of physics itself. Not only would consciousness become part of the world "out there," returning our science to some sort of alchemical status, but the premise of predictability would be (at least theoretically) invalidated.

The major philosophical implication of quantum mechanics is that there is no such thing as an independent observer. One of its founders, Werner Heisenberg, summarized this point in popular form in 1927 when he formulated his Uncertainty Principle. Imagine, he said, a microscope powerful enough to observe an atomic particle, such as an electron. We shine light down the instrument to enable observation, only to discover that the light possesses enough energy to knock the electron out of position. We can never see that particular electron, for the experiment itself alters its own results. Our consciousness, our behavior, becomes part of the experiment, and there is no clear boundary here between subject and object. We are sensuous participants in the very world we seek to describe.

In more technical terms, Heisenberg had discovered that position and momentum are complementary entities. One can determine the exact position of a particle only if one abandons the attempt to know anything about its motion (velocity), and vice versa. This discovery means that the Laplacian program is a delusion. Atomic or subatomic particles cannot be located precisely in space and time; and in an epistemology that equates the real with the material, the definition of the word "real" is suddenly open to question. Note that the Uncertainty Principle does *not* refer to a margin of error, which is present in every scientific experiment, and which reflects the accuracy of the verification of the prediction made. Instead, Heisenberg is talking about a probability that enters into the very definition of the state of the physical system. He says, in effect, that consciousness is part of the measurement and therefore reality (as it has been defined in the West for nearly four hundred years) is inherently blurry, or indeterminate.[13] The "change in the concept of reality manifesting itself in quantum theory," wrote Heisenberg in 1958, "is not simply a continuation of the past; it seems to be a real break in the structure of modern science." The so-called probability wave of quantum mechanics, he continued, "was a quantitative version of the old concept of 'potentia' in Aristotelian philosophy. It introduced something standing in the middle between the idea of an event and the actual event, a strange kind of physical reality just in the middle between possibility and reality." The break, of course, lies in the subject/object distinction itself; the "strange kind of physical reality" is consciousness, which we now see has material consequences. "What we observe," said Heisenberg, "is not nature in itself but nature exposed to our method of questioning." This was precisely Polanyi's point about tacit knowing. The great irony of quantum mechanics is that in the classic fashion of *yin* finally turning into *yang*, the Cartesian attempt to find the ultimate material entity, thereby "explaining" reality and ruling out subjectivity once and for all, resulted in discoveries that mocked Cartesian assumptions and established subjectivity as the cornerstone of "objective" knowledge.[14]

The enormous resistance of scientists to the philosophical implications of quantum mechanics is fully understandable, for once these implications are fully accepted, it becomes unclear just what is involved in "doing science." Either we are back to Aristotle's *potentia* (or the alchemical alembic), or we sit in a crowded stadium watching spoon-bending demonstrations by charlatans (but *are* they? That's the point!). Apparently, falling cubes *can* be influenced by mental concentration, and there is no way such information can be accomodated within the Cartesian paradigm.[15] Alternatively, quantum mechanics points to Buddhism and mysticism in its general scheme of the world, something first noted by Joseph Needham in *Science and Civilization in China*, and since elaborated upon by a number of writers.[16] The animism implicit in quantum mechanics has been explored *mathematically* by the physicist Evan Harris Walker, who argues that every particle in the universe possesses consciousness.[17] At the very least, we are forced to conclude that the "world" is not independent of "us." It is not composed of building blocks of matter, and indeed, exactly what matter is has become highly problematical. Everything, it seems, is related to everything else. The "lesson of modern physics is that the subject (perceiving apparatus) and object (the reality measured) form one seamless whole."[18] *Panta rhei*, said Heraclitus; everything flows, only process is real.

Quantum mechanics thus affords us a glimpse of a new participating consciousness, one that is not a simple reversion to naïve animism. As we consider the implications of quantum mechanics, it becomes quite clear that the most significant alteration of our scientific world view would stem from the deliberate inclusion in our scientific thinking of the awareness that we participate reality. Historically, we have been limited to a choice of two possibilities. One either asserted the existence of a disembodied intellect, as we have done since 1600A.D.; or one argued (contrary to what we manifestly perceive with our present consciousness) that stones, houses, furniture, clouds, this book and the ink in it are alive, possess an indwelling spirit—as men and women did believe prior to the Scientific Revolution. From what has been said above it should be clear that no matter how long the dominant culture continues to hold on to the first choice, that choice has no philosophical future. Both the discoveries of quantum mechanics and the Polanyi/Barfield analysis demonstrate that the totality of human consciousness, including tacit knowing and the information stored in the unconscious, is a significant factor in our perception and construction of reality. Like our X-ray student or ornithologist, we participate that reality subliminally in the learning process, and it later hardens into formulas that we then figurate as abstract entities. There is no need to make an external mystery out of this process, but it is an *internal* mystery, at least at this point in our understanding of the workings of the human mind. We have only the vaguest notion of how the conscious/unconscious interface operates, or how it brings us to conclusions about "reality." But since this thing, this alleged neuronal behavior pattern, operates partly in nonempirical ways (e.g., dreams, body knowledge), we are forced to conclude that the empirical/rational/mechanical view of nature, by denying nonempirical reality even while it depends on it, limits itself to descriptions of alpha-thoughts and conscious constructs. Such a view is thus both self-contradictory and erroneous. It must be supplemented so as to include our unconscious, to include nonempirical reality and the type of dialectical reasoning discussed in Chapter 3. But "supplemented" suggests the unintegrated addition of a lesser item and is thus a potentially misleading word. Perhaps the relationship I am suggesting can best be expressed by the metaphor of a nucleus embedded in a cell. The ego is embedded in a larger consciousness in which we participate, and acts as the organizer of life, and as in the cell, the proper relationship between the two modalities is osmotic. Modern science, on the other hand, identifies ego-knowledge with the whole of knowing; it tries to make that osmotic membrane rigid and impermeable. As a result, this type of consciousness begins to suffocate and die.

As it turns out, a number of thinkers are beginning to argue that the intellect, or conscious mind, is a subsystem of a larger system that we might call Mind with a capital M. This Mind is in fact the "strange kind of physical reality" of which Heisenberg spoke (above), suspended between possibility and reality. As Gregory Bateson has put it:

> The individual mind is immanent but not only in the body. It is immanent also in the pathways and messages outside the body; and there is a larger Mind of which the individual mind is only a subsystem. This larger Mind is comparable to God and is perhaps what some people mean by "God," but it is still immanent in the total interconnected social system and planetary ecology.[19]

There is no "transcendance" in this conceptual schema; there is no "God" present in the usual sense of the term. It is not *mana* that alters (or permeates) matter, but the human unconscious, or more comprehensively, Mind. There are no spirits out there within the rocks or trees, but neither is my relationship to those "objects" one of a disembodied intellect confronting inert items. My relationship to those "objects" is systemic, ecological in the broadest sense. The reality lies in my relationship with them. Just as two lovers create a relationship that is itself a particular entity (process), so does my working at the typewriter in front of me constitute an entity (process) that is larger than either Berman or Olympia Portable. My typewriter is not alive, there is no *original* participation here, but I am engaged with it in a process—writing this book, in fact—which is its own reality, and which is larger than either myself or the typewriter. The machine and I form a system so long as I engage its use or attend to its existence. As a result, the common perception of my skin as a sharp boundary between myself and the rest of the world begins to weaken, but without my becoming a schizophrenic or a preconscious infant.[20] A science that attends to such relationships rather than to so-called discrete entities would be a science of what has been called "participant observation," and it is this type of holistic thinking which might hold the key to future human evolution. This approach might qualify, in Ferenczi's words, as an "animism no longer anthropomorphic."

It should be clear that there is an enormous similarity between what Bateson is suggesting and the view of nature which emerges from quantum mechanics. Both state that it is inherent in the configuration of the relationship between ourselves and nature (to use the misleading language of Cartesian dichotomy) that we can never get more than a partial description of reality, or even of our own minds. Quantum mechanics implies that nature is *fundamentally* indeterministic, that elementary particles are *ontologically* always in partially defined states.[21] From this point of view, a direct correlation can be drawn between the mind/body dichotomy and the Freudian program of attempting to render the unconscious conscious. Bateson underlies the impossibility of what Freud wanted to do when he compares it to the attempt to construct "a television set which would report upon its screen *all* the workings of its component parts, including especially those parts concerned in this reporting."[22] It turns out that the subject/object distinction of modern science, the mind/body dichotomy of Descartes, and the conscious/unconscious distinction made by Freud, are all aspects of the same paradigm; they all involve the attempt to know what cannot, in principle, be known. The subject/object merger intrinsic to quantum mechanics, on the other hand, is part of a very different paradigm that involves a new mind/body, conscious/unconscious relationship. This mental framework, as both Bateson and Wilhelm Reich realized without making explicit, is similar to that of quantum mechanics in that it conceives of the relationship between the mind and the body as a field, alternately diaphanous and solid. In Wolfgang Pauli's terms, it "would be the more satisfactory solution if mind and body could be interpreted as complementary aspects of the same reality."[23] "There is no specific border in which mind becomes matter," writes philosopher Peter Koestenbaum; the "area of connection is more like a gradually thickening fog." There is no object existing by itself; every object has a stream of consciousness, or what we have called Mind, attached to it.[24]

This discussion brings us, finally, to Kant's *Ding an sich*, the inaccessible material substrate that supposedly underlies all phenomenal appearances. As Norman O. Brown has correctly pointed out, the flaw in Kant's system, and in all reasoning of this kind, is the equation of the categories of thought (space, time, causation) with human rationality—an equation that leads to the conviction that Mind and intellect are one and the same thing. Given the link between "us" and "nature" discussed above, the *Ding an sich* turns out to be the unconscious mind.[25] As Freud recognized, it is this mind that underlies all conscious awareness, and that pushes its way into consciousness when we manage to relax our ever vigilant repression. Once we recognize this situation, we must acknowledge that the question of the *Ding an sich* in nature is a red herring in exactly the same way as was the question "What was the alchemist actually doing? That there is *something* material out there, existing independently of us, would be useless to deny; that we are in a systemic or ecological relationship with it, unknowingly permeate and alter it with our own unconscious, and thus find in it what we seek, should be equally useless to deny. The future of "nature" itself thus depends on the recognition of the relationship between our own conscious and unconscious minds, and on what we do with that recognition.[26] In a post-Cartesian mode of thinking, "in here" and "out there" will cease to be separate categories and thus, as in an alchemical context, will cease to make sense. If we are in an ecological, systemic, permeable relationship with the "natural world," then we necessarily investigate "that world" when we explore what is in the "human unconscious," and vice versa.[27] Kant's *Ding an sich* is thus no longer unknowable. It is, however, never *fully* knowable, not *immediately* knowable, and it changes over time anyway. Note that this conceptual position does not reestablish naïve animism and it does not, in some fashionable, anti-intellectual sense, close down the enterprise of science. Instead, it opens the possibility of a new science, a larger one, a vista that, like the contemporary picture of the universe, is at once bounded but infinite.[28]

To summarize point (2), a systemic or ecological approach to nature would have as its premise the inclusion of the knower in the known. It would entail an official rejection of the present nonparticipating ideology, and an acceptance of the notion that we investigate not a collection of discrete entities confronting our minds (Minds), but the *relationship* between what has up to now been called "subject" and "object." One can draw an analogy between this notion and the field concept in electrodynamics, in which matter and force are seen as a system, and in which the energy resides in the field. A neo-holistic science would include ourselves in the force field. In its world view, the "energy" would reside in the relationship, or the formal (dynamic) ecology of the structure itself. The study of "nature" would thus be the study of "ourselves," and also the study of that force field. Stones do not fall to earth because of immanent purpose, and their rate of acceleration can certainly be measured by Galilean or Newtonian methods; but that behavior itself (i.e., our measurement of it) is conditioned by various forms of tacit knowing. The falling stone, the earth, and the Mind that participates this event from a relationship, and this, not some "spirit" in the stone or some "rate of acceleration," would be the subject of scientific inquiry.

Let us finally turn to point (3), the problem of radical relativism, which can be summarized as follows: the scientific method seems to discover laws and facts that are incontrovertible—gravity, equations governing projectile motion, the elliptical orbits of the planets. However, a historical analysis reveals that the method, and thus the findings, constitute the ideological aspect of a social and economic process unique to early modern Europe. If, as Karl Mannheim held, all knowledge is "situation-bound," it becomes difficult for any conceptual system, science included, to argue that it possesses an epistemological superiority over any other such system. Thus I argued in Chapter 2 that we must try to see science as a thought system adequate to a certain historical epoch, and attempt to separate ourselves from the common impression that it is some sort of absolute, transcultural truth. The implication is that there is no fixed reality, no underlying truth, but only relative truth, knowledge adequate to the circumstances that generated it. We see, then, that an analysis of science itself, using the method of the historical or social sciences, puts the validity of the scientific enterprise on an insecure footing. To make matters worse, it even undermines the historical analysis that precipitated this conclusion.

How can *any* conceptual system avoid such a paradoxical, and in fact self-destructive, result? It seems to me that in order to do so, a successful epistemology would have to be able to demonstrate the existence of an inherent truth or order in the conjunction between man and nature, and to survive the test of self-analysis. In other words, the application of its method to the method itself would not attenuate its validity.

Viewing radical relativism as we have just done, we are confronted with a remarkable realization: it is a problem for modern scientific epistemology alone. Radical relativism was *born* with the scientific method; it does not exist in any nonscientific culture or context. There is no such thing as a teleological analysis of Aristotelianism, a Hermetic analysis of alchemy, a quantum-mechanical analysis of quantum mechanics, or an artistic analysis of art (artistic and literary criticism are a mode of scientific explication, not themselves art or fiction). An artistic analysis of art, for example, could only involve deliberate parody: Dada, Andy Warhol, the *nouveau roman* or "anti-novel," and so on, but there are very sharp limits to these genres; they are really curiosities, and tend to have fairly short histories. Only modern science and its social and behavioral derivatives have this peculiar "creased" or "diptych" structure, whereby the discipline folds back on itself. One can put Freud on the analyst's couch, or discuss a mode of sociological analysis as being itself the product of certain social conditions, but one cannot possibly interpret the Aristotelian corpus as potentiality turning into actuality, or put the alchemist into his own alembic (he was ideally there already). This situation should not be confused with the "self-corrective" ability of modern science, which, as Polanyi demonstrates elsewhere in his book, does not really exist anyway.[29] As Karl Mannheim valiantly tried not to see all his life, this "diptych" structure is not self-correction, but self-destruction. It leads to philosophical paradoxes that were certainly known in antiquity, but formulated in the spirit of riddles or "brain-teasers." In modern times the sociology of knowledge, *a fortiori* the paradoxes it leads to, puts science and its derivatives on a shaky foundation—as Kurt Gödel, the discoverer of science's most famous paradox, found out.[30]

Why should this be the case? What does science lack that it falls prey to this problem? In a word, it lacks participation, or rather, the admission that it does involve participating consciousness. I know of no *logical* way to demonstrate that the denial of participation is the cause of radical relativism, and I am not advancing a causal argument of that sort; but they do seem to exhibit an observable pattern of interdependence. Modern science uniquely denies participation and uniquely has the problem of radical relativism, and it seems to me that it would be hard to have one without the other. Our earlier analysis suggests that participation is the "inherent truth or order in the conjunction between man and nature," and thus that the denial of participation must go hand in hand with convoluted thought patterns. As the case of quantum mechanics shows, modern epistemology is literally bursting at the seams from what it has tried to push out of conscious awareness. The attempt to equate conscious, empirical reality with the whole of reality is a futile task, for the unconscious will not be kept down. Once human subjectivity, tacit knowing, figuration, or whatever one wishes to call nonanimistic participation, is included in the thing known, the problem evaporates. Any system that acknowledges participating consciousness loses the "power" to analyze itself, because participation of whatever sort is the inclusion of the knower in the known. Effectively, then, the system already includes self-analysis as part of its method. Only if one shoves the self, the participant, out of the picture does one find oneself in the rather strange position of having that subjective entity, in schizophrenic fashion, float outside the creation and point out that the picture is seriously flawed.

Science, wrote Nietzsche in *The Birth of Tragedy*,

> spurred on by its energetic notions, approaches irresistibly those outer limits where the optimism implicit in logic must collapse. . . . When the inquirer, having pushed to the circumference, realizes how logic in that place curls about itself and bites its own tail, he is struck with a new kind of perception: a tragic perception.[31]

Or, as he says at another point in the same essay, "a culture built on scientific principles must perish once it admits illogic. . . ." Personally, I do not believe that a scientific culture such as ours, having run its course, analyzed itself, and discovered its limitations, has only tragedy or destruction to look forward to. Some collapse is inevitable, but this is not to say that destruction is necessarily the end point of it all. It is equally possible to face the error of nonparticipating consciousness squarely, and to begin the work of creating a new culture, one based on a new view of nature and a new scientific question. Nietzsche had the misfortune to draw his conclusions in an age when no respectable alternatives to scientific materialism were possible, and it is only under such conditions that tragedy or collapse is inevitable. We are not so delimited. The next step in the creation of a post-Cartesian paradigm, it would seem, is to place participating consciousness on a firm biological basis, that is, to demonstrate in physiological terms the existence of an "inherent truth or order in the conjunction between man and nature."

We have seen that science alone claims to be value-free even while it adheres to "objectivity" as a value; that the attempted separation of fact and value which characterized the Cartesian epoch can never be a serious philosophical possibility. Yet up to this point, our discussion has itself been purely abstract, disembodied. If an inherent order exists, it must be affective, because man is an emotional as well as an ideational entity. All of this suggests that a correct world view would have to be, at root, visceral/mimetic/sensuous. After four centuries of repression, Eros is finally coming in again through the back door.

6

Eros Regained

The flute of interior time is played whether we
   hear it or not.
What we mean by "love" is its sound coming in.
When love hits the farthest edge of excess, it reaches
   a wisdom.
And the fragrance of that knowledge!
It penetrates our thick bodies,
it goes through walls—
Its network of notes has a structure as if a million
   suns were arranged inside.
This tune has truth in it.
Where else have you heard a sound like this?

       —Kabir, fifteenth century, version by Robert Bly

Energy is the only life and is from the Body, and Reason is the
bound or outward circumference of Energy. . . . Energy is Eternal
Delight.

       —William Blake, *The Marriage of Heaven and Hell* (1793)

There is another world, but it is in this one.

       —Paul Eluard

153

There is, then, something missing from the analysis presented in the last chapter. Polanyi only hints at the importance of the body in the configuration of tacit knowing. He states that the latter is biological in nature, and that it has continuity with the knowledge possessed by children and animals. Yet this theme is never developed. Caught in the Cartesianism he rejects, Polanyi is not able to establish firmly the link between the visceral and the cerebral. To do that, one must be quite clear about rejecting the Cartesian paradigm while accepting the consequences that such a rejection entails. More significantly, one must be willing to *live* those consequences; and in a Cartesian culture, that is not an easy task.

Until recently, only two major scientific figures had met this challenge, and it is perhaps not an accident that they were both psychiatrists, immersed in the problem of how various individuals negotiated the boundary between "in here" and "out there." We

155

have discussed the first, Carl Jung, in some detail. As we saw, Jung broke with scientism, but doing so propelled him backward in time. In medieval and Renaissance alchemy he recognized a wholeness that permeated the psyche of the Middle Ages, and which was still present in human dream life. Clearly, dream analysis has a timeless importance, but any science constructed on Jungian premises would necessarily be a straightforward revival of the occult world view and thus a return to naive animism. Jung shows us the path to a *non*-Cartesian world view, but his premises cannot be the basis for a *post*-Cartesian paradigm, which this book seeks to define.

The second major scientific figure who lived the consequences of rejecting Cartesianism was Wilhelm Reich, and despite the unlikely claims and outright scientism of his later years, Reich's work is a major breakthrough in our knowledge of the mind/body relationship and an enormous contribution to any post-Cartesian epistemology. Since Reich, unlike Jung, was forward-looking (i.e., contemporary and politically progressive) rather than medieval in outlook, the social reaction to him could not be confined simply to branding him an obscurantist. That Reich is (to my knowledge) the only thinker to have had the distinction of seeing his works burned by the FBI suggests that he struck a fairly deep nerve and tends, in fact, to validate his own argument about the dialectical longing for, and hatred of, repressed instincts in Western industrial society. Reich attempted to reintroduce Dionysus to a culture gone berserk from Apollo, but the real importance of his work is that it points to the primacy of visceral understanding: the recognition that the intellect is grounded in affect, and the contention that instinctual repression is not merely unhealthy, but productive of a world view that is factually inaccurate. For our own purposes, Reich's work, specifically his understanding of the human unconscious, puts flesh and blood into the concept of tacit knowing, and in doing so, makes nonanimistic participation possible. With the scientific discovery that the body and the unconscious are one, and the concomitant recognition of a close relationship between the unconscious and tacit knowing, the subject/object distinction collapses, for body knowledge (sensual knowledge) then becomes a part of all cognition. The divorce between Logos and Eros may have been relatively brief, and these traditional partners in the search for truth may now be beginning to renegotiate their relationship.

Reich's discovery has remarkable implications for the whole question of participating consciousness. Since the seventeenth century only scientific thought has been regarded as truly cognitive; other types of understanding are "merely" emotion. The identity of the sensual and the intellectual was, as I have shown, the crux of the mimetic tradition, and is perhaps best illustrated by the decidedly nonmetaphorical use of the word "know" in the Bible: "And Abraham knew his wife Sarah." In the modern period, the relationship between science and other forms of knowledge or belief remains highly problematic. All serious philosophies that have made concessions to nondiscursive thought, notes Susanne Langer, have turned to mysticism or irrationalism, that is, "dispensed with thought altogether."[1] If Eros can be revived at all, it has to be through the claim that Eros is a fully articulated way of knowing the world, the ignorance of which has been intellectually crippling. It is precisely this that Reich, and his followers, have claimed.

In this chapter I hope first to demonstrate that the union of Eros and Logos is a scientific fact rooted in the experience of preconscious infancy, and thus that the holistic world view, or participating consciousness, has a physiological basis. Second, I wish to elaborate Reich's equation of the body with the unconscious and apply it to the concept of tacit knowing, thus making the point that the holistic experience of infancy continues to permeate adult cognition and understanding of the world. Taken together, these two points substantiate the analysis of Chapter 5 in a biological way; close the Cartesian paradigm down as a legitimate way of knowing reality; and open the door to an exploration of what might constitute a neo-holistic science.

Since Freud's first formulation of the subject, students of child development have largely agreed that the first three months of life constitute a period of "primary narcissism," or in Erich Neumann's terminology, the "cosmic-anonymous phase." The infant is all Unconscious (primary process) during this time, its life essentially a continuation of the intrauterine period. It behaves as though it and its mother were a dual unity, having a common boundary, and it lives as easily in others as in itself. External sensations, including the mother's breast, are perceived as coming from the inside. The world is largely explored by hands and mouth. "The child," writes Sam Keen in *Apology for Wonder*,

is, at first, a mouth, and his oral incorporation of the breast of the mother and other objects in the environment forms his initial way of relating to the exterior world. He quite literally tastes reality and tests it to see whether it is palatable. What promises delight to the taste buds—whether it be the breast, the thumb, or a nearby toy—he seeks to incorporate, to intuit, to take into himself.

For the infant, subject and object are almost completely undifferentiated, a fact that led Freud to argue that it was this particular perception that broke through adult dualistic consciousness in the mystic experience (Romain Rolland called this phenomenon the "oceanic feeling" in a letter he wrote Freud in 1927). The pleasure of reality is identical to the knowledge of reality at this point; fact and value are one and the same thing. "The surface of the body with its erogenous zones," writes Erich Neumann, "is the principle scene of the child's experience both of itself and of others; that is, the infant still experiences everything in its own skin."[2]

Between this stage and the child's third year, a gradual series of developments finally produces a discontinuity that constitutes the crystallization of the ego. Yet despite the birth trauma, the comparative harshness of modern child-rearing practices, and the inevitable frustrations of the environment, the term "cosmic anonymity" is not an inappropriate description of *all* of the first two postnatal years, a virtual paradise compared to what comes after. From the fetal period on, the infant body, or Unconscious, is subjected to the constant message of subject/object merger, of lack of tensions (and thus distinctions) between self and other. The enormous power of this message, which is the foundation of all holistic cognition, becomes apparent when we translate it into physiological terms. It means that the infant's entire existence is sensual, infinitely more sensual than it will ever be again. In Freud's famous formulation, the preconscious infant is "polymorphously perverse." More precisely, it is polymorphously whole. The entire surface of its body is an agent of sense, and its relationship to its surroundings almost completely tactile. Its entire body, and thus its entire world, is sensualized. For more than two full years, then, a fundamental realization is fostered in the body, or unconscious mind, of all of us, a foundation that can never be uprooted: *I am my environment.* Hence the phrase "*primary* process": the unconscious knowledge of the world, with its dreamlike structure of reasoning and cognition, comes first. The ego, Freud argued, is a secondary phenomenon; it is a structure that crystallizes *out of* cosmic anonymity.[3]

This situation raises an obvious question: Why leave the Garden of Eden at all? Why does ego-crystallization occur in the first place? Ego psychologists such as Margaret Mahler, Edith Jacobson, and Jean Piaget have dealt with this development as though it were an inherent and universal process. Freud, with his keen historical awareness, was not so easily misled. As our earlier discussion of the history of consciousness reveals, there was a time in human history when the ego did not crystallize out. Pre-Homeric man was completely, or almost completely, primary process, and his mode of knowing correspondingly mimetic. Throughout the Middle Ages people saw themselves as continuous with the environment to some degree, the alchemist being the chief spokesman for this perception. As we saw, the final break occurred only towards the end of the sixteenth century; that is really what *Don Quixote* is all about. Being aware that ego-crystallization in general was a relatively recent development, Freud resolved the problem of its emergence in the individual with the phylogeny/ontogeny argument that the growth of the modern infant recapitulated the history of the race as a whole. But if we accept this formulation, and do not see ego-development as at least partly innate, we must then argue (as Freud did for most of his career) that the ego is forced to crystallize out as a result of the frustrating impact of reality (i.e., the environment). Hence his expression "reality principle," and his famous dictum, "Where id is, there shall ego be." But this statement, if true, implies that reality, especially in the form of child-rearing practices, must have become increasingly frustrating with the passage of centuries, and that there must have been some sort of turning point at the end of the Middle Ages, when ego-strength made its appearance in full-blown form. In fact, ego-development does have its innate aspects, but is also a cultural artifact: there does seem to be a history of increasing alienation that climaxed on the eve of the Scientific Revolution.

Before discussing the innate and learned (historical) aspects of ego-development, however, I wish to emphasize the staggering implications of the previous paragraph. If Freud's line of reasoning is correct, then the ego, which we take for granted as a given of normal human life, is not only just a cultural artifact, but—in its contemporary form, at least—actually a product of the capitalist, or industrial, epoch. The quality of ego-strength, which modern society regards as a yardstick of mental health, is a mode of being-in-the-world which is fully "natural" only since the Renaissance. In reality, it is merely adaptive, a tool necessary for functioning in a

manipulative and reifying (i.e., life-denying) society. This historically conditioned nature of ego also suggests that if modern society
in its present form were to disappear, "man" as we understand
him would vanish as well—a rather eerie conclusion that Michel
Foucault was unable to avoid in the concluding pages of *The Order
of Things*. In other words, a different way of life might not only
mean the end of ego-strength as a virtue, but of ego-strength as a
way of existing, and therefore of "man" as he is currently conceived. Equally surprising (perhaps) is the implication that what
we regard as healthy personality traits are the product of attitudes
toward children, and child-rearing practices, that are hopelessly
neurotic—a thesis central to Reichian psychology.[4]

To take the issue of ego-development first, then, recent research
has shown that the first two years of life, even the first three
months, are not as anonymous or unconscious as Freud and
Neumann believed. Newborn infants can localize a touch on the
skin, or a source of sound, though not with any great accuracy.
They can locate the position of an object in space, and begin to
imitate at six days of age. If the mother sticks out her tongue, so
will the baby, and as Thomas Bower points out, this is a complex
achievement. The baby recognizes that its own tongue (which it
can only know by the feel of it) matches its mother's tongue, which
it can see. This identification of its own body parts with those of
others is a primitive form of subject/object correlation.[5]

At about four or five months, the unspecific smile characteristic
of the first three months becomes a particular response to the
mother. The child acquires a new look of alertness and attentive-
ness; it is no longer drifting. For Margaret Mahler, this shift in
perception is the onset of body-ego formation. At six months, the
baby begins to experiment, pulling at the mother's hair or face,
putting food into her mouth, straining away from her to get a
better look. At seven or eight months, the pattern of comparative
scanning begins. The child looks away from the mother and back to
her, comparing the familiar with the unfamiliar. At age eight
months, the child begins to distinguish between different objects,
between father and mother, for example, and also to respond to
facial indicators of mood. At nine months children no longer automatically grab at anything presented, but first stop to look at
what is being offered. The belief in object constancy, that an object
continues to exist when not in view, develops within the next three
months.[6]

160

Other aspects of ego-development can be seen by charting a child's behavior in front of the mirror. The first awareness of one's body-image in the mirror occurs at about six months, at which time the child smiles at the image of another. From six to eight months, it begins to slow down its movements in front of the mirror and start relating them to the movement of the image, appearing thoughtful as it does so. At nine to ten months, it makes deliberate movements while observing its image, actually experimenting with the relationship between itself and the image. At twelve months, the child recognizes that the image is a symbol, but its grasp of that fact remains precarious for a while, and thus it continues to play with its reflection, in some cases up to thirty-one months of age.[7]

From ten/twelve months to sixteen/eighteen months the child begins to practice with its larger environment. It moves away from the mother physically by crawling (but still holding on, occasionally); eventually, it masters upright locomotion. The child now begins to perceive mother from a greater distance and establish familiarity with a wider segment of the world. From fifteen to twenty-four months the original "cosmic unity" starts to break down in earnest. The child begins to balance separation and re-union by "shadowing" the mother (watching and following), then darting away, expecting to be chased and picked up. There is both a wish for reunion and a fear of reengulfment. The mother is now a person in its mind, not just "home base." The little boy or girl starts to bring things back from the outside world to show her. He or she also begins to experience the body as a personal possession, not wishing it to be handled. The child learns to cope with mother's absences, and develops disappearance/reappearance games. It will practice deliberately hiding toys and then finding them, or standing in front of a mirror and suddenly ducking out of sight. Mother or father will be instructed to cover their own eyes ("don't see me") and then abruptly uncover them ("see me"), or told to pretend not to see the child and then suddenly "discover" it with exaggerated glee. Language develops in the second year of life, emerging out of a "babbling" phase in which the child makes all kinds of sounds, both invented and imitated. The use of the pronoun "I" occurs at about twenty-one months.[8]

So innate do all these actions appear that it would seem impossible to argue for a two-year period in which primary process is dominant. A nascent ego seems to be present and growing from birth. Yet we have to ask ourselves what we mean by ego, or ego-consciousness. Clearly, the pre-Homeric Greeks, who did not possess such consciousness, went through many of the processes just described, including the evolution of a brilliantly sophisticated language system. All of these developments may be necessary conditions for ego-crystallization, but they are not sufficient ones. Ego-consciousness can in fact be compared to pregnancy. There are degrees of it, but (to quote an old saw) one cannot be "just a little bit pregnant." Like a quantum jump, ego-consciousness involves a specific kind of discontinuity, and in the modern infant it occurs at roughly two and a half years of age, when the child one day has the startling thought: "I am I." The child begins to use the pronoun "I" several months before this event, we should add; it is no surprise that it exists in pre-Homeric Greek and all ancient languages. But this is not the same thing as having the thought, "I am I." The latter expresses a wholly different level of existence, one involving the recognition that ultimately you cannot be known by the other and are radically separate from him. This recognition takes place at about the same time that the child becomes convinced of what its image in the mirror represents, and is, as Merleau-Ponty notes, the beginning of alienation. From this point on the child begins to recognize that it is visible for others, and that there is a conflict between the "me" it feels and the "me" others see. The outside world, the child now realizes, can interpret it in a way that denies its own experience of itself. The third year of life (at least in modern Western cultures) is thus a trying period for parents as the child goes about establishing its identity with a determined stubbornness. Indeed, failure to be a "bad boy" or girl at this point can result in eventual psychosis, the key fear being that you are totally transparent to others, being nothing more than what they interpret you to be. The healthy child often objects to being watched at this time, for it now understands that its identity goes beyond the roles or the situation it is in, that it is an "I," an ego at odds with the world (to some extent), and to the interpretation that the world might place upon it. Dualistic consciousness is now an irrevocable fact.[9]

We should thus not confuse motor and perceptual skills with ego-crystallization per se, for as we saw in Chapter 3 (following the analysis of Julian Jaynes), entire civilizations can be built without benefit of the latter. One can generate governments and wars, construct the ziggurat or the Code of Hammurabi, and even predict eclipses, without benefit of an ego. In order to undertake such

projects one certainly has to be able to imitate, grasp, and locate objects in space, but they require no soul-searching or self-
awareness. I belabor this point because it is so difficult for us, with
our own ego-consciousness, to comprehend that ego-
crystallization is a comparatively recent development; that one can
move through all or most of the stages of motor and perceptual
development described above without ego-discontinuity taking
place. At most, then, one can say that ego-development is partly
innate, but that it apparently requires certain cultural triggers to
"spring" it, to tip the balance all the way. Whereas ego-
crystallization may be natural, it does not follow that it is inevitable. Furthermore, the range of ego-strengths present in the world
today, especially from culture to culture, as well as the gradual
hardening that occurred between Plato's Greece and the Scientific
Revolution (with a strong upswing thereafter), shows that even
within the context of ego-discontinuity a great variety of behavior
is possible. All the evidence thus points to the limits of ego psychology, which, through its laboratory experiments with children,
tries to establish a case for the innateness and universality of ego-
crystallization.

What exactly *is* the ego, then? Although they are not the same
thing, ego and language possess important structural similarities.
As Daniel Yankelovich and William Barrett point out in their
pioneering study, *Ego and Instinct,* ego and language are the joint
product of evolution and culture, and their development will not
take place if society does not provide critical experiences at the
appropriate time. If the "babbling" phase of language does not
occur in a social context, the child will not learn to speak at all—as
has been documented in a few cases where children were discovered being raised by animal species. Both languages and ego can be
regarded as "incomplete psychic structures," or what the authors
call *developmentals:* "structures that grow only when phylogenetic
factors interact with critical individual experience at specific stages
in the life cycle." Such individual experience, however, is really
social in nature, and it varies significantly from culture to culture
and between different historical epochs. [10]

The recognition that cultural factors are important for ego-
crystallization actually lurks in the survey of supposedly innate ego-development presented above. As Thomas Bower
points out, certain perceptions are innate and certain ones acquired. [11] Not all cultures believe in object constancy or solidity, for

example, nor is it clear that children of every culture practice "shadowing" games with the mother, or games of "see me"/"don't see me," or those of third-year identity-testing. In earlier times, such games were probably absent altogether. Ego-strength is much softer in nonindustrial cultures than in ours, and such ego-developments are probably correspondingly weaker. Studies such as the one Gregory Bateson and Margaret Mead did of Bali, for example, reveal child-rearing patterns that have little in common with our own (see Chapter 7). In a similar vein, the objection to body handling that occurs at around eighteen months of age was not present during the Middle Ages and is apparently still absent in many Third World cultures. [12]

In contrast, we discover that some of the mothers in Margaret Mahler's study (see note 2) were highly motivated by the prestige attached to being part of a research unit at the Masters Children's Center in New York and, as a result, they were often achievement-oriented with regard to their children, wanting them to be as precocious as possible in their sensorimotor development. Both researchers and mothers watched anxiously for signs of ego-development (or what they took to be such signs). Had these not arisen in any particular child, it would have been branded autistic. Yet at some point in the history of the race, we were all "autistic," and it was ego-development that was viewed with alarm. The strong contemporary bias in favor of ego-development cannot help but prejudice the "scientific" study of it. The research unit at Masters was, in fact, a perfect mirror of the American ethos. The classic Jewish joke, "my son the doctor" (aged six months), is not just a *Jewish* joke; it is the norm for Western industrial societies, which turn out rigid ego-structures with a vengeance. It becomes difficult to demarcate sharply between innate and acquired when the infant is subject to a socialization process that begins with its first breath. [13]

Though the issue of which cultural factors trigger ego-crystallization is immensely complex, and (since ego is erroneously regarded as a universal human characteristic) very poorly researched, one factor can be pointed to with some degree of certainty. It is quite clear that the history of increasing ego-development in the West is also the history of increasing repression and erotic deprivation, manifested over the centuries by a drop in the body contact and sensual enjoyment that normally occurs during the first two years of life. Ego-development is not merely purchased at the expense of sensual enjoyment (the classical theory of sublimation); more significantly, it has repression (i.e., sexual alienation) as a condition necessary—and possibly even sufficient—for its development. In short, enough repression may tip the balance, and "impregnate" the psyche. Let us briefly examine the evidence for this thesis.

Prior to the rise of agricultural civilization (i.e., before ca. 8,000 B.C.), man lived as a hunter-gatherer. Of necessity, mothers carried their babies on the body almost all the time. Mother and child were not separated after birth. They slept together, and the mother breast-fed the child for nearly four years. Feeding was dependent on spontaneous hunger rather than prearranged schedules.[14]

Much of this practice was retained in the ensuing millennia. Nursing in ancient Judea, for example, averaged two to three years, and babies were still carried around, rather than put in a crib or left unattended. Older children were taken on the shoulder or carried astride, as is still the custom in Third World cultures. The Greeks typically transferred the neonate to a basin of warm water, to maintain the continuity of intrauterine experience. In the eleventh century A.D., the great Arab physician Avicenna recommended nursing for two years, and urged gradual rather than sudden weaning—caveats that may suggest the existence of some departure from the custom of extended breast-feeding.[15]

The significance of breast-feeding, curiously enough, lies less in the chemical value of the mother's milk than in the cutaneous stimulation provided by the accompanying maternal-infant contact. In *Touching: The Human Significance of the Skin,* Ashley Montagu gathered mountains of evidence to show that in all mammalian species, a healthy adult life is not possible without a large amount of tactile stimulation during the first few years, and especially the first few months, after birth. Indeed, the proper development of the nervous system, including myelinization (formation of the fatty sheath of protective tissue around the nerves), depends on it. Although the quantity of tactile stimulation of infants has tapered off over time, it was maintained to a very great extent down to about 1500 A.D. Whether through direct carrying, extended nursing, or even the gentle manipulation of the infant's genitals, body stimulation was a large part of early life, and all of these practices are still maintained in those parts of the world which are as yet unaffected by modernization.[16]

Direct correlations cannot be made, but child-rearing practices among contemporary non-Western cultures may be indicative of what was typical in the West down to the early Renaissance. In Bali, for example, the child is carried on the hip or in a sling, in almost constant contact with the mother for the first two years of life. During the first six months it is never *not* in someone's arms except while being bathed, and the parents typically play with the male infant's genitals when he is in the bath. Similar information has been gathered about a number of contemporary "primitive" societies, and the matter of playing with the infant's genitals was singled out as a point of comparison by Philippe Ariès in *Centuries of Childhood*. In the Middle Ages, he tells us, public physical contact with children's private parts was an amusing sort of game, forbidden only when the child reached puberty. This attitude changed sharply during the Renaissance, but is, Ariès notes, still widespread in Islamic cultures. Interestingly enough, practices such as placing the neonate in a warm bath, or encouraging infantile sexuality, are slowly making something of a comeback, the rationale being that such practices lead to a less anxious and more healthy sexual life. [17]

Ariès also provides a detailed study of late medieval attitudes toward children, which imply that this was a period of changing practices in the matter of body contact. [18] Indeed, the single most important theme of his book is separation, dissociation. Ariès is able to show that prior to the late sixteenth century, neither the nuclear family nor the child existed as *concepts*. Until the twelfth century, art did not portray the morphology of childhood, and portraits of children were almost nonexistent until the end of the sixteenth century. The seventeenth century literally "discovered" childhood, and made a point of demarcating it as a stage in a series of separate phases of life. Far from implying greater care of infants, however, this demarcation involved greater alienation from them. Special children's clothing was now used to make visible the stages of growth and, at the end of the sixteenth century, there suddenly emerged a great preoccupation with the supposed dangers of touching and body contact. Children were taught to conceal their bodies from others. In addition, it was now believed that children must never be left alone. The result was that the adult became a sort of psychic watchdog, always supervising the child but never fondling it—a practice that is really the prototype of scientific observation and experimentation.

These same patterns were institutionalized in the colleges of the late Middle Ages, where they took the form of constant supervision, a system of informing (i.e., spying), and the extended application of corporal punishment. The birch replaced fines as the predominant penalty, and students were commonly whipped in public until they bled. By the eighteenth century, flogging occurred on a daily basis in England, where it was viewed as a way of teaching children and adolescents self-control.

The late Middle Ages thus saw an abrupt shift in the emphasis of child-rearing practices, a shift from nurturing to mastery which was one aspect of the emergence of a civilization marked by categorization and control. As child-rearing practices reveal, Western society was still heavily sexualized down to the sixteenth century. It was "the essentially masculine civilization of modern times," as Ariès puts it, which discouraged such nurturing practices. The rise of the nuclear family, with the man at the head, reached full expression in the seventeenth century, whereas the crucial unit had previously been the "line," that is, the extended family of descendants from a single ancestor. With the evolution of the nuclear unit, the soft heterogeneity of communal life began to disappear. Distinctions were made within the family and between families. The medieval household, which might contain up to thirty members of the extended family, began to shrink and become uniform. Beds, which used to be scattered everywhere, were now confined to a special room. What we would call chaos was in effect the multiplicity of realities, a "medley of colors," says Ariès, and it is still observable in the streets of (say) Delhi or Benares, where eight types of transport and forty different types of people can be seen on a single narrow street, or in the throngs of people which crowd the streets of Mediterranean towns after sunset. "Masculine" civilization, with its desire to have everything neat, clean, and uniform, erupted in full force on the eve of the Scientific Revolution. From the thirteenth century onward, the power of the wife steadily diminished, the law of primogeniture (the eldest son has exclusive right of inheritance) being a prime example of this. Down to the mid-sixteenth century, no man save the occasional astrologer was allowed to be present when a woman gave birth. By 1700, a very great percentage of "midwives" was male. "Professional" civilization, the world of categorization and control, is a world of male power and dominance.

The desensualization of childhood, and the subsumption of child-rearing under masculine control and scientific management, reached their apogee in the twentieth century. This development has not, of course, been without its positive consequences. We cannot, for example, ignore the marked drop in the infant mortality rate. But the accompanying psychic cost of this desensualization may lead us to question how much has really been gained. I am not referring here to child *abuse*, which has apparently declined over the centuries, but to desexualization, estrangement, to being "out of touch," a condition that arises when the parent relates to the child with a deliberate failure of responsiveness. Abusive treatment can be as sexual as loving treatment, and it is anything but unresponsive.[19] It may create angry adults, but it does not of itself lead to existential anxiety. It is the latter condition that comprises the daily fare of today's adult; and it is crucial to note that this same existential anxiety characterizes the consciousness of the schizoid personality, which, according to Ashley Montagu, can itself often result from a lack of tactile stimulation in infancy.[20] And given the assembly line of modern obstetrics, this situation is perhaps no surprise. How does the child enter the world in Western industrial societies? "The moment it is born," writes Montagu,

> the cord is cut or clamped, the child is exhibited to its mother, and then it is taken away by a nurse to a babyroom called the nursery, so called presumably because the one thing that is not done in it is the nursing of the baby. Here it is weighed, measured, its physical and any other traits recorded, a number is put around its wrist, and it is then put in a crib to howl away to its heart's discontent.

The child is put on a fixed feeding schedule that is maintained for months, and which has little relation to its own hunger pangs. Rapid weaning from the breast is encouraged by modern medicine, if indeed the child is breast-fed at all.

That cutaneous stimulation is crucial for health, if not life itself, is not difficult to demonstrate. During the nineteenth century more than half the infants in the United States died in the first year of life from marasmus, a word that literally means "wasting away." As late as 1920, the death rate for this age group in foundling institutions, where absolutely no body contact was provided, was *nearly 100 percent*. As Montagu explains, American infant care was then under the influence of Luther Emmett Holt, Sr., a professor of pediatrics and the Dr. Spock of his generation, whose popular writings urged fixed feeding schedules, abolition of the cradle, and a minimum of fondling. J. B. Watson, the founder of behavioral psychology, was also very influential at this time, and he urged mothers to keep their emotional distance from their children. He specifically stated that such treatment, in addition to fixed feeding schedules, strict regimens, and toilet training, would mold the child's capacities in a manner that would facilitate its conquest of the world. The goal, he said, was to make the child "as free as possible of sensitivities to people"—an objective that has, in the late twentieth century, come to fruition with stunning "success."[21]

Though it may be difficult to make a strict causal argument here (a fact that continues to plague modern anthropology[22]), it is noteworthy that the discarding of the cradle, the abandonment of fondling, and the rise of mechanistic child-rearing practices have gained ground in those Third World countries that have taken industrial development and Westernization as their express purpose. It is somehow understood that science, "progress," and dehumanized child-rearing practices go hand in hand. The formula becomes, to turn E. M. Forster upside down, "only disconnect."

Further evidence for the destructive influence of modern child-rearing practices has been provided by Marshall Klaus and John Kennell of the Case Western Reserve School of Medicine in Cleveland. Their studies reveal that when the birth is natural and not interfered with by the institution, there is a common pattern to mother-infant bonding. The first sixty to ninety minutes of life are an extraordinary period, during which the neonate is unusually alert and engages with the mother in a sort of primeval bonding "dance," in which the two touch, fondle, and gaze profoundly into each other's eyes. The modern hospital does not permit this interaction to occur, however. The mother is often given painkillers, which dull her perception, and medication is routinely applied to the newborn's eyes, blurring its vision. In fact these practices make no real difference, for the hospitals immediately separate mother and child, with quite noticeable effects. In one experiment, Klaus and Kennell compared a group of mother-infant pairs that were allowed sixteen hours of immediate contact to a control group that was not. Two years later, the mothers in the first group dealt with their children in a relaxed way, using more questions and adjectives, and fewer commands, in their speech. The second set was caught up in scolding, inhibiting, and giving frequent commands. Sixteen hours of fondling apparently had an effect lasting two years. Klaus and Kennell also visited nurseries in Guatemala, where there is extensive early body contact between mothers and children, and witnessed much less fussing and crying. Similar variations in behavior were observed by Louis Sander and his colleagues at the Boston University Medical Center. They found that babies raised by nurses were affected adversely if the nurses' orientation was markedly "professional," that is, geared to the hospital staff rather than to the children.[23]

What is the implication of this survey of child-rearing practices for ego-crystallization? Although no causal connections can be confidently asserted, it does seem that there is a historical gestalt at work. Simply put, contemporary "primitive" cultures, similar to the West before 1600, have much softer ego-structures than we do, and are characterized by a more communal and heterogeneous way of life, far less anxiety and madness, and much gentler subject/object distinctions. In general, says Montagu, adult personalities in extended body-contact cultures are less competitive; and those few "primitive" societies that do not have such contact, such as the Mundugumor people of New Guinea studied by Margaret Mead, produce irritable and anxious adults.[24] These findings are hardly surprising. Child-rearing in Western industrial culture is so stark that it is not difficult to understand that it is crucial in the maintenance, if not the genesis, of modern anomie. Reich's sadomasochism, Laing's schizoid personality, Sartre's nausea are conditions that could thrive only in such a desexualized context.

Of course, the ego has its positive aspects. It certainly existed in the West from about 800 B.C. to 1600 A.D. without massive alienation as its corollary, but it is hard to avoid the conclusion that in its modern form the ego is the product and expression of pathology. Specifically, it seems to be (again, in its modern form) a structure evolved to obtain love by way of mastery in an unloving world. But as Reich pointed out, love and mastery are, physiologically, incompatible goals. We search desperately for love and authenticity, but in the context of a world that has taught us to fear these very things. The results are, inevitably, mass neurosis and substitute gratification (see Plate 17). In a curious parody of the Uncertainty Principle, the very precision of the modern ego has created a kind of parataxis in our social relations, whereby they seem to be foggy, disconnected, even autistic. This is the tragic message of the Beatles' *Sgt. Pepper* album, released in 1967, essentially a set of vignettes about human dissociation. "Will you still need me, will you still feed me, / When I'm sixty four" could well be the "national anthem" of the industrialized world.[25]

170

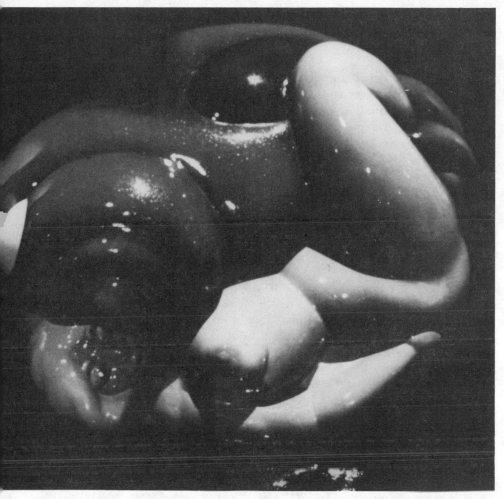

Plate 17. Luis Jimenez, Jr., *The American Dream* (1969/76). Fiberglass and epoxy, 20″ × 35″ × 30″. By permission of the artist.

The sickness of contemporary life, pervaded as it is with heavy drug use and alcoholism, stems from the futile attempt of scientific culture to eradicate holistic perception. But holistic cognition is a primary, ecological perception of nature rooted in a biological substrate, and present before the ego ever arises. The history of archaic man, and the cosmic-anonymous phase of childhood, bear clear witness to the existence of this primeval substrate of primary-process material. This stratum is hardly a developmental; it is the ground of our being, and unlike the ego, does not need

171

cultural factors to trigger it. No amount of civilization can eradicate it, and the scientific attempt to do so can only drive us to drink. We never escape the impact of the cosmic-anonymous phase; participation remains the basis of our perception throughout our lives. "The primary unitary reality," writes Erich Neumann in *The Child*, "is not merely something that precedes our experience; it remains the foundation of our existence even after our consciousness, grown independent with the separation of the systems, has begun to elaborate its scientifically objective view of the world."[26]

Holism haunts modern man, tugs unmercifully at his consciousness. Despite the way he is forced to live, he still hears that preconscious echo, "I am my environment." He is trained into asceticism, writes Norman O. Brown, trained into a posture of analytical distancing from nature, yet he remains unconvinced, "because in infancy he tasted the fruit of the tree of life, and knows that it is good, and never forgets."[27] As Reich realized, this memory is stored in the body, and whether expressed in the terms of original participation (the occult world view), or through the deliberate resexualization of life (which Reich courageously attempted to effect), there is no getting away from it.[28] It is for this reason that primary-process material is at the root of all premodern epistemologies, that children's thought patterns are largely magical in structure down to about age seven, and that participating consciousness survives, even in modern scientific epistemology. What the child, the "primitive," and the madman know, and the average adult fights to keep out of his or her conscious awareness, is that the skin is an artificial boundary; that self and other really do merge in some unspecified way. In the last analysis, we cannot avoid the conviction that everything really *is* related to everything else.

In effect, it was this continuity of holistic perception which Reich sought to demonstrate in scientific (and later, scientistic) terms. To do so one must show that unconscious knowledge is essentially body knowledge or, more plainly put, that the body and the unconscious are one and the same thing, and this was precisely Reich's major contribution to psychoanalysis. A brief sketch of his work will help to substantiate our argument for the continuity of holistic consciousness.

Freud, as is well known, adhered religiously to the Cartesian paradigm. For him, as for Descartes, all affect was ultimately rooted in the mechanical arrangement of corpuscles (or neurons), a belief made explicit in his unpublished "Scientific Project" of 1895

172

and retained by Western medicine to this day. Mind and body, or ego and instinct, are rigidly distinct entities, and all intrapsychic processes (like everything else) are essentially mechanical in nature. From this strictly materialistic analysis, with its elaboration in terms of thermodynamic and hydraulic energy transfers (conversion, cathexis, resistance, etc.), it followed that neurotic symptoms were adventitious, or mechanically separable. In other words, a neurosis for Freud was an alien element in an otherwise healthy organism. It was formed by repressing a painful event and thereby removing it from conscious awareness; the neurosis could itself be removed by techniques (notably free association) designed to make the unconscious memory conscious.

As thousands of Freudian analysts and analysands have come to realize, this sensible, intellectual approach does not work. Freud himself was aware of its limitations and did emphasize that the therapy session must flush out, or "abreact," the emotion that accompanied the original repression. Yet his commitment was ultimately to the supposedly curative power of the intellect. "I can only wonder what neurotics will do in the future," he remarked naïvely to Jung, "when all their symbols have been unmasked. It will then be impossible to have a neurosis."[29] That analytical cognition made little difference for affect, or that *mimesis* might be knowledge, were notions that Freud was no more willing to accept than Plato had been. Nor did he ever grasp how passionately, even erotically, he was attached to the concept of intellectual knowing.

Reich, like Jung, was keenly aware of the limitations of this approach. His central argument was that what we call "personality," or "character," was itself a neurosis; or, as psychiatrist John Bowlby has put it, a posture of defense against the threat of object-loss. Against Freud's mechanistic theory, with its idea of separable parts, Reich advanced a holistic one: "*there cannot be a neurotic symptom,*" he wrote, "*without a disturbance of the character as a whole.* Symptoms are merely peaks on the mountain ridge which the neurotic character represents."[30]

The "mountain ridge" to which Reich referred is the specific structure of the personality, which has a psychic aspect, the neurosis, and a muscular one, the character armor. Early in life, he contended, the spontaneous nature of the child is subjected to severe repression by its parents, who fear such spontaneity (in particular, the lack of sexual and sensual inhibition) and socialize it out of the child, as it was long ago socialized out of them. By age

173

four or five, the natural instincts have been crushed or surrounded by a psychic defence structure that has a muscular rigidity as its correlate. What is lost is the ability to succumb to involuntary experience, to abandon control and lose oneself in an activity; to obtain what Reich called (perhaps misleadingly) "orgastic gratification." The orgastically ungratified person develops an artificial character and a fear of spontaneity. Whereas the healthy character is in control of his or her armor, the neurotic character is controlled *by* it. The emotions of the latter, including anger, anxiety, sexual desire, or whatever, are rigidly held down by this muscular tension, and the result is the stiff (or collapsed) posture and mechanical articulation of the body that is observable almost everywhere in our society. This neurotic character, or "modal personality,"[31] encased in character armor, might most appropriately be compared to a crustacean. Its entire character is designed to fulfill the function of defense and protection or, alternatively, acquisition and aggrandizement. It moves from crisis to crisis, driven by a desire for success and proud of his ability to tolerate stress. Its armoring is not merely a defense against the other, but against its own unconscious, its own body. The armor may protect against pain and anger, but it also protects against everything else. These emotions are held down by inverted values, such as compulsive morality and social politeness—the veneer of civilization. The modal personality is thus a mixture of external conformity and internal rebellion. It reproduces, like a sheep, the ideology of the society that molded it in the first place, and thus its ideology (regardless of its politics) is essentially life-negating. In reproducing that ideology, the neurotic character produces its own suppression. Neurosis is not some adventitious accretion, some fly in the ointment. It is, Reich argued, an icon of personality and culture as a whole.

We have already met the modal personality of the modern era in Isaac Newton, and have noted the relationship between his self-repression and his system of the world. We have also argued that such a person was the product of the rise of capitalism and the Puritan mentality that accompanied it. In one of his earliest studies, Erich Fromm demonstrated quite convincingly the connection between this so-called anal type, with its preoccupation with orderliness, and the social typology of the capitalist described by Werner Sombart and Max Weber. "The character structure," wrote Reich, "is the congealed sociological process of a given epoch." As Reich realized, such a type is hardly the prerogative of capitalist

174

society, for it exists in all industrial societies, all societies based on production and efficiency rather than joy and authenticity.[32]

How does one cure such a person; which is to say, most of us? Reich had a strong political orientation, and did not believe that individual cures could succeed apart from major social changes. But the project of integrating individual with social change eluded him (as it has every political theorist), and he was not able to clarify how a political program could be forged out of authenticity or self-realization. On the individual level, however, he had no doubts: authenticity meant, specifically, body authenticity, the feeling of the continuity of consciousness with the body which Descartes denied was possible. "The philosophic underpinning of body authenticity," writes Peter Koestenbaum, "is that the body is a metaphor for the fundamental structure of being itself"—a position, incidentally, with which no self-respecting alchemist would disagree.[33] The restoration of authenticity, of the sense of authentic being-in-the-world, was thus not likely to be accomplished through the intellect; a situation that for Reich explained the general failure of Freudian analysis. Reich's specific mode of therapy went hand in hand with his realization that Descartes was quite simply wrong, that the mind/body dichotomy was an artificial construct. The whole theory of character armor, which Reich believed was validated every time a patient walked into his office, demonstrated that "muscular attitudes and character attitudes have the same function in the psychic mechanism." The psychiatrist could actually have greater success in getting to the unconscious through the manipulation of the patient's body than by the technique of free association. This manipulation loosened the armor, producing not merely an abundance of twitchings and sensations, but primitive emotions and a memory of the event during which these emotions (instincts) were originally repressed. These emotions and memories were not, in Cartesian formulation, causes or results of body phenomena; rather, "they were simply these phenomena themselves in the somatic realm." Somatic rigidity, wrote Reich, "represents the most essential part in the process of repression," and each rigidity "*contains the history and meaning of its origin.*" Armor, in short, is the form in which the experience of impaired functioning is preserved. Reich concluded not only that the traditional mind/body dichotomy was in error, but that Freud was wrong in arguing that the unconscious, like Kant's *Ding an sich*, was not tangible. *Put your hands on the body*, said Reich, *and you*

175

have put your hands on the unconscious. The eruption of ancient childhood memories and their affective accompaniment in hundreds of patients demonstrated to him that the unconscious can be contacted directly in the form of the biological energy of the body and the various twists and turns that have blocked and distorted it.

The identity of the body and the unconscious, which Reich was able to demonstrate clinically, is something we are all intuitively aware of, and which can be explored without undergoing Reichian analysis. All of us have had the experience, for example, of waking up and forgetting what we were just dreaming about. We may then slowly shift our position in bed, only to have part or all of the dream come back; and different positions will retrieve different scenes of the dream. In dreaming, apparently, certain imagery from the body tissues is released as we toss and turn in our sleep; or alternatively, these images got "fixed" in the body while it was in certain positions. Recalling a particular image is therefore often dependent on assuming the bodily configuration that was present during the original dream sequence.

Reich's insights have profound implications for epistemology. The Cartesian mind/body split diagramed in Chapter 1 is in reality the schema of the modern schizoid personality. This personality can also be schematized as in Figure 12. What we take as normal is thus a distortion of a very different, non-Cartesian relationship that a person can and should have with him- or herself, as illustrated in Figure 13.

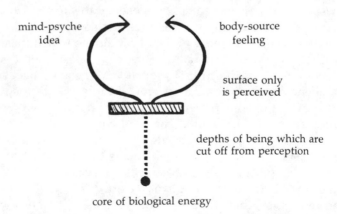

Figure 12. Wilhelm Reich's schema of the neurotic personality (from Alexander Lowen, *Depression and the Body*, p. 304).

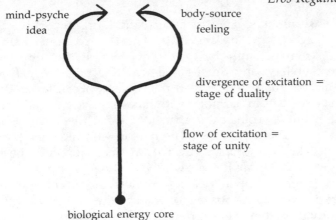

mind-psyche
idea

body-source
feeling

divergence of excitation =
stage of duality

flow of excitation =
stage of unity

biological energy core

Figure 13. Wilhelm Reich's schema of the healthy personality (from Alexander Lowen, *Depression and the Body*, p. 303).

Since the Cartesian or Newtonian personality sees *only* duality, *only* subject/object distinction, the stage of unity indicated in Figure 13 is permanently inaccessible to him or her. But as we have seen, this unity is the primary reality of all human being and cognition, and to be out of touch with it is to be suffering from severe internal distortion. The point is that the modal personality, having a distorted internal relationship, must necessarily have a distorted external one. He or she will see the world the way Newton saw it in his later years. Surface appearances will be confused with the real thing. Truly accurate perception depends upon maintaining contact with the biological core, for only then can one return to it at will, that is, abandon control and merge with the object. And it was this ability to surrender control, to obtain "orgastic gratification," or what I have referred to as the mimetic experience, that Reich defined as the ability to love. Suspension of ego thus lies at the core of loving, and all true experience of nature depends on it.

The "secret" that lies at the heart of the occult world view, with its sense of everything being alive and interrelated, is that the world is sensual at its core; that this is the essence of reality. Tactile experience can be taken as the root metaphor for *mimesis* in general. When the Indian does a rain dance, for example, he is not assuming an automatic response. There is no failed technology here, rather, he is inviting the clouds to join him, to respond to the invocation. He is, in effect, asking to make love to them, and like any normal lover they may or may not be in the mood. *This is the way nature works.* By means of this approach, the native learns about the reality of the situation, the moods of the earth and the

177

skies. He surrenders: *mimesis*, participation, orgastic gratification. Western technology, on the other hand, seeds the clouds by airplane. It takes nature by force, "masters" it, has no time for mood or subtlety, and thus, along with the rain, we get noise, pollution, and the potential disruption of the ozone layer. Rather than put ourselves in harmony with nature, we seek to conquer it, and the result is ecological destruction. Who, then, knows more about nature, about "reality"? The person who caresses it, or the one who takes it by force, vexes it, as Bacon urged? The epistemological corollary of Reich's work is that having certainty about reality is dependent upon loving—a remarkable sort of conclusion. Conversely, perception based on mechanical causality and the mind/body dichotomy is best put under the heading of "impaired reality-testing," the clinical definition of insanity.

I do not mean to imply that primary process is somehow "good" and ego-consciousness correspondingly "bad," or that they are distinct, unrelated entities. Such an implication does, unhappily, lurk in Reich's writings. He did seem to believe, like Rousseau, that natural man was hidden under social man. The problem is that although primary process is the substrate, the ground of being, it seems clear enough that once the ego is triggered, it is, like a tree, as real as the soil from which it sprang. As in the case of language, learned and instinctual aspects here form a complicated and interrelated pattern. Reich's position must thus be modified to square with the theory of developmentals, which rightly argues against arbitrary distinctions between the instinctual and the acquired.

According to Yankelovich and Barrett, a great number of ethologists have concluded that although certain types of behavior—breathing, sucking, eating, sexual activity—develop independently of any culture, there is no behavior that does not display some aspects of learning. Even cells do not develop independently, but go through chains of environmental reactions with neighboring cells. No single instance of behavior has allowed a scientist to say, "this is pure instinct," without another researcher being able to demonstrate traces of learning in that same case. Since we have no infallible rules for distinguishing between innate and acquired, the best way to view the developmentals, say the authors, is as entities (or processes) in which experience and instinct are "regarded as inseparable aspects of a single unified event."[34] Although primary process is, as the phrase indicates, primary, we are forced to conclude that *both mimesis (identification) and analysis (discrimination) are present within the physiological response system of the human organism.* Since this conclusion holds even if one element, or process, is more fundamental than another, my critique of the ego has not been directed against the ego per se, but against the particularly virulent form that has, since 1600, insisted on a rigid mind/body, subject/object dichotomy. Prior to the Renaissance, the ego coexisted with participation more than it sought to deny it, and this attitude is what made it a viable structure for so many centuries. In denying participation, however, the ego denies its own source, for as both Reich and Freud (for most of his life) contended, the ego has no separate energy reserves of its own. The unconscious is the ground of its being. Like the nucleus of a cell, the ego is a contractile point within the Mind, and the Mind is the sum of knowledge gained by *all* of the body, all of the senses. In recognizing this position of the brain within the Mind, one biomedical engineer has suggested that the brain is not the source of thought but a thought amplifier; that knowledge originates not in the brain but in the body, and that the brain simply magnifies and organizes it. This thesis does not mean that the brain's processing function is somehow alien to the human physiological response system, any more than the nucleus can be regarded as an alien element in a cell.[35] Hence, the issue is not whether *mimesis* is good and analysis bad, but how and to what extent a given culture triggers the latter, that is, what it produces as the ecology of its typical personality. The culture of archaic man, through social attitudes, body contact, spontaneous feeding, and so on, hardly triggered the ego at all, *if* at all; "advanced" industrial societies seem to trigger nothing else. It may be the case, as Foucault suggests, that we shall reverse that trend and eventually return to a completely mimetic state; but it is not my contention (as it may have been Reich's) that such a consciousness would be the best that the human race could have, and, in any event, it is hardly an option we can act on. The ego, far more than modern science, is a part of our cultural baggage, so much so that to talk of deliberately "eradicating" it does not make much sense. At present, our only visible option is to modify it, and so go beyond it.

We are now in a position to give Polanyi's analysis of knowledge a biological underpinning. Given Reich's clinical identification of the body with the unconscious, our discussion of participation, figuration, and Polanyi's "tacit knowing" takes on a whole new dimension. Although Polanyi argued, in *Personal Knowledge*, that

179

such knowing was physiological, he was never able to prove his point, to establish that connection. *Reich supplies that missing link.* For if the body and the unconscious are the same thing, the permeation of nature by the latter explains why participation still exists, why sensual knowledge is a part of all cognition, and why the admission of this situation is not a return to primitive animism. It also explains why "objective" knowledge does not exist, and why all true knowledge (as Polanyi argued) constitutes a commitment. Taken together, Reich and Polanyi point the way out of the Cartesian paradigm, and into Ferenczi's "erotic sense of reality."

Let me try to state this another way, before elaborating the argument. That non-discursive knowledge has cognitive content may be a little-known fact in our culture, but it is hardly an *unknown* fact. Should the reader pick up Reich's *Character Analysis*, Albert Scheflen's *How Behavior Means*, Rudolf Arnheim's *Visual Thinking*, Susanne Langer's *Feeling and Form*, Andrew Greeley's *Ecstasy: A Way of Knowing*, or any of Freud's or Jung's works on dream symbolism, he or she will discover, in essence, a common theme. Nor do these few works, selected more or less at random, exhaust the topic. Since the late nineteenth century, a significant number of Western intellectuals have come to grips with the limitations of verbal-rational knowledge and have devoted their lives to demonstrating the different cognitive schema present in art, dreams, the body, fantasy, and illusion. What they have *not* succeeded in doing is showing the relationship between these two forms of knowledge. As a result, they have unwittingly exacerbated the "two cultures" split, a trend that is currently being reinforced with the popular dichotomy between "right-brained" and "left-brained" thinking.[36] If we are ever to break free of the Cartesian paradigm, we must do more than simply delineate the contours of nondiscursive knowing; we must show how the two forms of knowledge relate to one another. As long as they remain two cultures, or two brains, the dominant culture or brain can continue to take itself seriously while sanctimoniously paying lip service to the other. Reich's work, as well as that of Polanyi and Barfield, takes the first step toward a synthesis, for it demonstrates that the Cartesian paradigm is actually a fraud: there *is* no such thing as purely discursive knowing, and the sickness of our time is not the absence of participation but the stubborn denial that it exists—the denial of the body and its role in our cognition of reality.

What, then, is that role? What might a Reichian interpretation of Polanyi look like, modified by the theory of developmentals? Polanyi argued, first, that attributing truth to any methodology, scientific or otherwise, is a nonrational commitment, an act of faith, an affective statement. Second, he demonstrated that most of the knowing that we do is actually unconscious, or what he calls "tacit." The learning takes place by doing, in bicycle riding, language acquisition, or X-ray pathology. Our awareness of the underlying rules is subliminal, picked up by osmosis. There is nothing that is initially cognitive or analytical about the learning process, despite what we like to think. From a Reichian standpoint, the crucial issue is that commitment, and noncognitive comprehension of reality, are mimetic; they come about through identification, or collapse of subject/object distinction. Polanyi's paradigm case, the example of X-ray pathology, demonstrated this point quite dramatically. The X rays began to take on meaning as the student forgot his self and instead submerged his whole being in the experience.

What Reich would argue here, of course, is that participated knowledge is sensual. It is the *body* that is making the commitment in this study of X rays, that is absorbing the sights, sounds, and smells, and that has already incorporated the rules of the culture at large and is now doing the same with the subculture of X-ray pathology. We are literally back to the preconscious infant who knows the world by putting it in its mouth. Reality that is not "tasted" does not remain real to us. In order to make a thing real, we must go out to it with our bodies and absorb it with our bodies, for (as Hobbes once wrote) "there is no conception in man's mind, which hath not first been begotten upon the organs of Sense." It is only *after* this occurs, as I stated in Chapter 5, that rationality begins its work of reflecting on the information and establishing the categories of thought. It is at dusk, wrote Hegel, that the Owl of Minerva begins its flight, and that is why, except in the case of a scientific revolution, we wind up verifying the paradigm, finding out what we somehow knew all along.

The case of scientific revolution, where (as T. S. Kuhn argued) anomalies pile up so as to generate a crisis, is also more comprehensible on a Reichian interpretation than on a strictly intellectual one. If anomalies were nothing more than logical or empirical contradictions, we would never feel threatened by them. But when our world view is thrown into doubt, we feel anxiety, and anxiety is a visceral reaction. As Peter Marris shows in his book *Loss and Change,* all real loss involves grief and mourning, and the loss of a paradigm is often an emotional catastrophe. Marris, like Reich, supplies the visceral understanding lacking in Polanyi. Knowledge is learned, and generated, first and foremost by the body, and it is the body that suffers when serious changes are required.[37]

In Reichian terms, Polanyi's tacit knowing can be reformulated as follows. The *Ding an sich* in nature is the *Ding an sich* in ourselves, namely our bodies, or unconscious minds, which can never be fully known. As long as we continue to have bodies, there will be tacit knowing. Such knowing permeates nature and our cognition of it; the primary unitary reality of preconscious infancy is never abandoned, and represents the inherent order in the conjunction of man and nature. The knower is thus fully included in the known. When we get to the smallest particles in the universe, we discover our own minds in them, or behind them.

Furthermore, as we become adults, our bodies become more than just primary process. The unconscious is not a static, unchanging "thing." The cultural paradigm of the age is fed into our tacit knowing and then shapes our conscious knowing. The gradual decision to view projectile motion as parabolic, for example, came many decades after cannon and long-range firing had become fixtures of the environment, along with the increasingly utilitarian climate generated by the advent of bookkeeping, surveying, and engineering. Galileo learned about projectiles in the same way Polanyi's medical student learned about X-ray pathology, but his unconscious already carried the gestalt of a new age that had been building for nearly three centuries. We recognize, then, that there exists a close relationship between the cultural and the biological. Learning to figurate reality according to the rules of a culture would seem to be a heavily biological process, for the world view apparently gets buried in the tissues of the body along with the primary unitary reality. Indeed, this close relationship between the cultural and the biological may be part of the reason that the shape of the human body has changed over the centuries. A different consciousness must mean a different body, or as Reich would have (more accurately) put it, a different consciousness *is* a different body.[38]

Finally, we can translate the discussion of Mind provided in Chapter 5 into visceral terms, for what I mean by "Mind" is the conjunction of the world and the body—*all* of the body, brain and ego functions included. Once Mind so defined is recognized as the way we confront the world, we realize that we no longer "confront" it. Like the alchemist, we permeate it, for we recognize that we are continuous with it. Only a disembodied intellect can confront "matter," "data," or "phenomena"—loaded terms that Western culture uses to maintain the subject/object distinction. With this latter paradigm discarded, we enter the world of sensual science, and leave Descartes behind once and for all. Whereas a medieval denial of participating consciousness would have amounted to a denial of ghosts and fairies, the Cartesian denial of it is quite simply a denial of the body, a denial that we even possess a body. But once the body is understood to be an instrument of knowledge, and its denial seen as constituting as much of an error as any of Bacon's famous "Idols," we have made sensual or affective science theoretically possible.[39]

In Chapter 5, I suggested that the systemic view of nature did not close down the enterprise of science but in fact opened it up, creating a whole new set of issues for us to explore. It seems to me that the notion of Mind, or system, discussed in that chapter, and interpreted in terms of the present chapter, lays the groundwork for a nonanimistic, participated reality. We must pursue this notion further, however, asking several questions that will help us to grasp it in greater detail. What, for example, would a holistic experiment consist of? What types of answers might a holistic science provide?

"In the last analysis," wrote E. A. Burtt in *The Metaphysical Foundations of Modern Science*, "it is the ultimate picture which an age forms of the nature of its world that is its most fundamental possession. It is the final controlling factor in all thinking whatever." In *Philosophy in a New Key*, Susanne Langer elaborates this theme by stating that the crucial changes in philosophy are not changes in the answers to traditional questions, but changes in the questions that are asked. "It is the mode of handling problems, rather than what they are about, that assigns them to an age." A new key in philosophy does not solve the old questions; it *rejects* them. The generative ideas of the seventeenth century, she says, notably the subject/object dichotomy of Descartes, have served their term, and their paradoxes now clog our thinking. "If we would have new knowledge," she concludes, "we must get us a whole world of new questions."[40]

Langer has articulated the essence of our problem. We do not

183

need a new solution to the mind/body problem, or a new way of viewing the subject/object relationship. We need to deny that such distinctions exist, and once done, to formulate a new set of scientific questions based on a new modality. When I studied physics in college, for example, a unit was devoted to heat, then to light, then to electricity and magnetism, and so on. The project involved in each unit, the "generative idea," was, in effect, to ascertain the nature of light, heat, electromagnetism, etc. We see in this curriculum the strong grip of the Cartesian paradigm. Fifty years after the formulation of quantum mechanics, these subjects are still taught as though there can be a knowledge of them independent of a human observer. Again, I am not taking a Berkeleyian position: whether these things exist independently of our observation of them is not something I regard as a fruitful line of inquiry. What *is* at issue is the notion that observation makes no difference for what we learn about the thing being investigated. It is by now abundantly clear that we are part of any experiment, that the act of investigation alters the knowledge obtained, and that given this situation, any attempt to know all of nature through a unit-by-unit analysis of its "components" is very much a delusion. A question such as "What is light?" can have only one answer in a post-Cartesian world: "That question has no meaning."

How should we study (i.e., participate) nature? What questions *should* we ask? The reader is aware that I am not a scientist and am probably the wrong person to try to answer these questions. But having started this discussion, I am obliged to make some attempt to finish it, hoping to provide some valuable suggestions that others might develop further. Since I have already dealt extensively with the study of light, let me continue to organize the discussion around this problem. My choice, of course, is not arbitrary, for Newton's study of the nature of light became the atomistic paradigm, the model of how all phenomenon should be examined. I am thus attempting to grasp, by working with an archetypal example, what a sensual or holistic science might become; what it would mean to acknowledge participation by deliberately including the knower in the known.[41]

We saw in Chapter 1 that Newton, in his prism experiments, was able to show that a beam of white light was composed of seven monochromatic rays, and that each color could be identified by a number, signifying the degree of refrangibility. Today, the significant number is taken to be wavelength or frequency, but the Newtonian definition of color as a number is fully preserved. Red, for example, is the sensation caused by such-and-such a wavelength of light in the eye of a standard observer.

The Newtonian theory of color received a serious jolt in the 1950s from the work of Edwin Land, the inventor of the Polaroid camera. Land was able to demonstrate that colors were not simply a matter of wavelength, but that their perception was largely dependent on the objects or images that they represented; in short, on their context and its (human) interpretation. A white vase bathed in blue light is seen as white because the mind (Mind) accepts whatever the general illumination is, as white. The same phenomenon can be seen in the case of yellow automobile lights or candle flames, which are commonly perceived as white. Land discovered that even two closely placed wavelengths of light, for instance two different shades of red, can generate the full range of color in the eye of the observer.

In trying to make sense of this clear-cut refutation of the classical theory of light and color, Land was led to an explanation that echoed the critique of Newton made by Goethe in his much-ridiculed book *Farbenlehre* (On the Theory of Colors, 1810). "The answer," wrote Land, "is that their work [i.e., the work of Newton and his followers] had very little to do with color as we normally see it." (Goethe's phrase was: "Derived phenomena should not be given first place.") In other words, superimposed rays of monochromatic light are artificially isolated in the laboratory, and although no one is denying their importance in (for example) laser technology, they simply do not occur in nature. In his own experiments, Land discovered that the characteristic arrangement of colors was indeed a spectrum, but one that ranged from warm to cool—something artists have known for centuries. "The important visual scale," he concluded "is not the Newtonian spectrum. For all its beauty the [Newtonian] spectrum is simply the accidental consequence of arranging stimuli in order of wavelength."

Of course, there is nothing accidental about this arrangement. The value system of Newton's Europe deemed it sensible to identify colors with numbers or to arrange them in order of wavelength. The perception of colors in atomistic, quantifiable terms was made possible by Western industrial culture and ultimately delivered back to this culture technological devices, such as the sodium vapor lamp or the spectroscope, that "verified" this perception in a beautifully circular way. More significant here is the

185

fact that Land's experiments demonstrate that the Newtonian spectrum is *one way* of looking at light and color, but that there is nothing holy about it. Furthermore, Land's conclusions reveal the repression implicit in Newtonian science, even in this one special case, for the talk of warm versus cool colors plunges us directly into affect, and into human subjective interpretation. Degrees of refrangibility are supposedly "out there," eternal, not requiring a human observer to establish their validity. Hot and cold, however, are "in here" as well as "out there"; they require a human *participant*, in particular, one with a body and its accompanying emotions. Nor are degrees of refrangibility very stimulating emotionally. The quantification of color represents a dramatic narrowing of emotional response. The linguist Benjamin Lee Whorf was fond of pointing out that eskimos have thirteen different words for white, and certain African tribes up to ninety words for green. In contrast, European languages collapse an entire range of emotion and observation into three or four words: for example, green, blue-green, aqua, turquoise. We begin to understand what Lao-tzu meant when he said, "the five colors will blind a man's sight."

In any holisitc experiment with light and color, then, the important thing is that affect and analysis not be differentiated. If the experiment does not include emotional/visceral responses, it is not scientific, and therefore not meaningful. This approach does *not* rule out the Newtonian color theory. The "validity" of the classical theory of color, however, lies not in something inherent in nature, but in our appreciation and enjoyment of it; and one can certainly enjoy lasers, spectroscopes, and games with prisms. But if this theory is going to exhaust the investigation of the subject, then it is unscientific by virtue of omission. Land's work may thus be seen as the beginning of a paradigm for the holistic investigation of light and color. In the same vein, research on the psychology of color had demonstrated that a red rectangle does feel warmer and larger than a blue one of equal size. Certain combinations of colors make us feel sad, euphoric, dizzy, or claustrophobic. A number of prisons in the United States have recently installed a "pink room," incarceration in which for a mere fifteen minutes reduces the victim, à la *Clockwork Orange,* to complete passivity.[42] Phrases such as "I feel blue" or "that makes me see red" are not just metaphors, and an entire discipline, called "chromo-therapy" by its practitioners, has grown up around the intuitive recognition that certain colors have healing properties. We also now know that a field of colors, called an "aura," surrounds every living thing, and that children perceive it up to a certain age. It is likely that auras are still commonly perceived in nonindustrial cultures, and probable that the yellow halos painted around the heads of various saints in medieval art were something actually seen, not (according to a modern formulation) a metaphor for holiness "tacked on" for religious effect.

All of this is by way of suggestion. I cannot formulate a new, fully articulated paradigm, but I believe that the holistic exploration of such inexhaustible subjects as color, heat, or electricity, will give us—as Susanne Langer urged—a whole new world of questions. The key scientific question must cease to be "What is light?," "What is electricity?," and become instead, "What is the *human experience* of light?" "What is the *human experience* of electricity?" The point is not simplistically to discard current knowledge of these subjects. Maxwell's equations and the Newton's spectrum are clearly part of the human experience. The point is instead to recognize the error that arises when the human experience is defined as that which occurs from the neck up—the "Idol of the Head," we might call it. It is the incompleteness of Cartesian science which has made its interpretation of nature so inaccurate. "What is the human experience of nature?" must become the rallying cry of a new subject/object-ivity.[43]

The late twentieth century may be a difficult time to be alive, but it is not without its exciting aspects. At the very point that the mechanical philosophy has played all its cards, and at which the Cartesian paradigm, in its attempt to know everything, has ironically exhausted the very mode of knowing which it represents, the door to a whole new world and way of life is slowly swinging ajar. What is dissolving is not the ego itself, but the ego-ridigity of the modern era, the "masculine civilization" identified by Ariès, or what the poet Robert Bly calls "father consciousness." We are witnessing the modification of this entity by a reemergent "mother consciousness," the mimetic/erotic view of nature (see Plate 18). "I write of mother consciousness," states Bly in his breathtaking essay, "I Came Out of the Mother Naked,"

> using a great deal of father consciousness. But there is no other possibility for a man. A man's father consciousness cannot be eradicated. If he tries that, he will lose everything. All he can hope to do is to join his father consciousness and his mother consciousness so as to experience what is beyond the father veil.

Plate 18. Donald Brodeur, *Eros Regained* (1975). By permission of the artist.

Right now we long to say that father consciousness is bad, and
mother consciousness is good. But we know it is father consciousness
saying that; it insists on putting labels on things. They are both good.
The Greeks and the Jews were right to pull away from the Mother and
drive on into father consciousness; and their forward movement gave
both cultures a marvelous luminosity. But now the turn has
come....[44]

It is noteworthy that Bly credits the nonparticipating consciousness of the Greeks and the Jews as producing cultures of "marvelous luminosity," for in doing so he poses a caveat for all thorough-
going Reichians. It may well be that the culture of Europe from the
Renaissance to the present has been based on sensual repression;
and Reich may well have been right in believing (unlike Freud) that
culture per se did not *have* to depend on repression; but whatever
the energy that fueled it, the brilliance of modern European culture

188

is surely beyond doubt. The whole of the Middle Ages did not produce a sculptor like Michaelangelo, a painter like Rembrandt, a writer like Shakespeare, or a scientist like Galileo; and in terms of sheer volume of creativity, the comparison is even more dramatic. Bly's crucial point, however, is that the "marvelous luminosity" has reached its limits. It has become a hostile glare, a scorching ball of fire that, as Dali tried to suggest, even melts clocks in an arid desert landscape. Its most creative outposts are now self-criticisms, analyses of the culture that double it back on itself; quantum mechanics, surrealist art, the works of James Joyce, T. S. Eliot, and Claude Lévi-Strauss. There is a chance, as Bly suggests, that a more luminous culture "lies beyond the father veil," one that may warm and nurture rather than burn and dessicate. Indeed, as an act of faith, I am convinced of it. But for now, it is clear that the sharp subject/object dualism of modern science, and the technological culture that religiously adheres to it, are grounded in a developmental gone awry. Cartesian dualism, and the science erected on its false premises, are by and large the cognitive expression of a profound biopsychic disturbance. Carried to their logical conclusion, they have finally come to represent the most unecological and self-destructive culture and personality type that the world has ever seen. The idea of mastery over nature, and of economic rationality, are but partial impulses in the human being which in modern times have become organizers of the whole of human life.[45] Regaining our health, and developing a more accurate epistemology, is not a matter of trying to destroy ego-consciousness, but rather, as Bly suggests, a process that must involve a merger of mother and father consciousness, or more precisely, of mimetic and cognitive knowing. It is for this reason that I regard contemporary attempts to create a holistic science as the great project, and the great drama, of the late twentieth century.

7

Tomorrow's Metaphysics (1)

Let me state my belief that such matters as the bilateral symmetry of an animal, the patterned arrangement of leaves in a plant, the escalation of an armaments race, the processes of courtship, the nature of play, the grammar of a sentence, the mystery of biological evolution, and the contemporary crises in man's relation to his environment, can only be understood in terms of such an ecology of ideas as I propose.

—Gregory Bateson, Introduction to
Steps to an Ecology of Mind (1972)

W e have come a long way since our survey of seventeenth-century science, and our analysis of the shift from feudalism to capitalism which accompanied the emergence of the Cartesian paradigm as the dominant world view of the West. I have argued that science became the integrating mythology of industrial society, and that because of the fundamental errors of that epistemology, the whole system is now dysfunctional, a mere two centuries after its implementation. A view of reality structured on what is only conscious and empirical, and excluding the tacit knowing that any perception in fact depends upon, has brought us to an impasse. I have suggested that the split between analysis and affect which characterizes modern science cannot be extended any further without the virtual end of the human race, and that our only hope is a very different sort of integrating mythology.

At the end of Chapter 6, I made some suggestions as to how fact and value might once again be united—suggestions that could possibly become part of a new epistemology, but which do not constitute a coherent system in and of themselves. There are, however, a large number of disciplines that claim to unite fact and value, and some of these, such as yoga, Zen, the oriental martial arts, and various types of meditation, are rapidly gaining popularity in the West. In addition, a number of well-articulated philosophies, such as those of George Gurdjieff and Rudolf Steiner, offer coherent, monistic ways of understanding the world. Why not adopt one of these? Why not abandon Cartesianism and embrace an outlook that is avowedly mystical and quasi-religious, that preserves the superior monistic insight that Cartesianism lacks? Why not deliberately return to alchemy, or animism, or number mysticism? If reality frightens you, Max Weber once remarked, the religion of your fathers is always there to welcome you back into its loving arms.

The problem with these mystical or occult philosophies is that they share what Susanne Langer has cited as the key problem of all nondiscursive thought systems: they wind up dispensing with thought altogether. To say this is not, however, to deny their wisdom. Such philosophies contain the nugget of participating consciousness and can make it real to any serious devotee, and for that reason alone, practices such as Zen and yoga are certainly worth doing. My point is that once the insight is obtained, then what? These systems are, like dreams, a royal road to the unconscious, and that is fine; but what of nature, and our relation to it? What of society, and our relationship to each other? If our goal is nothing more ambitious than calming our anxieties and turning off our minds—as is typically the case when an empire or major world view collapses—then we can simply turn philosophy into psychotherapy and be done with all these discomforting complexities. Intellectually, this approach is not very interesting, and psychologically, it strikes me as being a colossal failure of nerve. In fact, it is but the flip side of Cartesianism; whereas the latter ignores value, the former dispenses with fact. It seems to me that we should be able to do better than merely alternate between extremes.

In larger terms, the problem may be restated as follows. We stand at a crossroads in the evolution of Western consciousness. One fork retains all the assumptions of the Industrial Revolution and would lead us to salvation through science and technology; in short, it holds that the very paradigm that got us into trouble can somehow get us out. Its proponents (and they generally include the modern socialist states) view an expanding economy, increased urbanization, and cultural homogeneity on a Western model as both good and inevitable. The other fork leads to a future that is as yet somewhat obscure. Its advocates are an amorphous mass of Luddites, ecologists, regional separatists, steady-state economists, mystics, occultists, and pastoral romantics. Their goal is the preservation (or resuscitation) of such things as the natural environment, regional culture, archaic modes of thought, organic community structures, and highly decentralized political autonomy. The first fork clearly leads to a blind alley or Brave New World. The second, on the other hand, often appears to be a naïve attempt to turn around and go back whence we came; to return to the safety of a feudal age now gone by. But a crucial distinction must be introduced here: recapturing a reality is not the same thing as returning to it. My discussion of alchemy attempted to clarify how much we lost when that tradition was discarded. In Chapter 6, I sought to demonstrate that if one equated body knowledge with unconscious knowledge, the Hermetic world view became physiological rather than occult. But at no point did I suggest that we could solve our dilemmas by attempting to return to the premodern world. Rather, my point was that as long as we dream, and as long as we have bodies, the insight into reality which the alchemists, Jung, and Reich obtained will remain indispensable, and must in fact become a major part of our view of reality. The same thing can be said for the attempt to live in harmony with the environment, or to have a sense of intimacy and community. Such things will always be the basic reality of a healthy human life, and a world view that ignores them in the name of "progress" is itself a precarious illusion. "All the errors and follies of magic, religion, and mystical traditions," writes Philip Slater in *Earthwalk*, "are outweighed by the one great wisdom they contain—the awareness of humanity's organic embeddedness in a complex and natural system."[1] Recapturing this wisdom is not the same thing as abolishing modernity—although it might help us to transcend it.

The real difficulty, of course, is discovering how to recapture this wisdom in a mature form. The works of Jung and Reich are landmark attempts to do this, but their approach tends to be anti-intellectual. A knowledge of dreams and the body will inevitably be crucial components of the new metaphysics, but I doubt that the work of Jung and Reich could ever serve as its framework.

Indeed, I know only one attempt to reunite fact and value which I regard as a possible framework for a new metaphysics, and that is the astonishing synthesis provided by the cultural anthropologist Gregory Bateson. As far as I can tell, his work represents the only fully articulated holistic science available today; one that is both scientific *and* based on unconscious knowing. Bateson's work is also much broader than that of Jung or Reich, in that it places a strong emphasis on the social and natural environment, in addition to the unconscious mind. It situates us *in* the world, whereas Jungian or Reichian self-realization often becomes an attempt to avoid it.

Bateson is not yet widely known, but I suspect that future historians may come to regard him as the most seminal thinker of the twentieth century. The "Batesonian synthesis"—what might be termed the "cybernetic/biological metaphor"—is not Bateson's work alone; but the synthesis of ideas is his, as is the extraction of the concept of Mind from its traditionally religious context, and the demonstration that it is an element inherent in the real world. With Bateson's work, Mind (which also includes value) becomes a concrete reality and a working scientific concept. The resulting merger of fact and value represents an enormous challenge to the human spirit, not merely a calming of its fears.[2]

As we begin our discussion of Bateson, however, it may be useful to provide a disclaimer at the outset. Modern science got into trouble by claiming to be the one true description of reality. In this sense it had much in common with its predecessor, the medieval Catholic world view, and there is no point in deliberately repeating this error. I am not suggesting, then, that Bateson's work is without limits or problems, or that the crises of our age can be resolved simply by adopting it uncritically and applying it to our dilemmas. Crises don't get resolved that way in any event. What I do believe is that Bateson's work represents the recovery of the alchemical world view in a credible, scientific form; that it turns the conscious/unconscious dialectic into a creative method for investigating reality; and that if the world view of a nondystopian New Age is not derived directly from his work, it will inevitably contain some of its most salient features.

Although the Batesonian synthesis bears remarkable similarities to Eastern thought, and appears to be epistemologically disparate from all Western scientific methodology save quantum mechanics and information theory, its real inspiration was the work of Gregory's father, William Bateson, the remarkable turn-of-the-century biologist who coined the term "genetics" in 1906. A brief exposition of William Bateson's scientific career is indispensible not merely to an understanding of the origins of Gregory Bateson's thought, but also to a thorough grasp of its content.[3]

William Bateson lived in the heyday of British scientific materialism. The great physicist James Clerk Maxwell (1831–79) had published his final statement on reality, *Matter and Motion*, towards the end of his life, and Thomas Henry Huxley had spent much of his career popularizing that way of thinking in his ideological "campaign" for physical science and the Darwinian theory of natural selection. Bateson, who had received his own training under the famous anti-Darwinian thinker Samuel Butler, was, despite his scientific sophistication, part of an older nonprofessional scientific tradition, that of the "gentleman amateur," a social type closely associated with the British aristocracy.[4] Materialism, utilitarianism, and expertise—all these he saw as the shoddy values of a bourgeois middle class. His own emphasis was on aesthetic sensibility. He spoke of true education as "the awakening to ecstasy" (an idea retained by Gregory in his own theory of learning), not the dreary preparation for a mundane career. Scientific work reached its highest point, he held, when it aspired to art. As an undergraduate at Cambridge he defended the retention of classical Greek as a required subject because it provided an "oasis of reverence" in the otherwise arid mind of the typical science student, and in an 1891 flyer on the subject, he wrote:

> If there had been no poets there would have been no problems, for surely the unlettered scientist of to-day would never have found them. *To him it is easier to solve a difficulty than to feel it.* [Italics mine]

Creating a science out of the "feel" of things proved an accomplishment that eluded William Bateson. His own career embodied the agonizing split between science and art, the healing of which became the central project of Gregory Bateson's life. He was convinced from the start that emotion, like reason, had precise algorithms, and one of Gregory's favorite quotes was taken from Descartes' arch-rival, Blaise Pascal: "The heart has its *reasons* which the reason does not at all perceive."

William Bateson's attempt to create a science of form and pattern, and the aesthetic and political attitudes that formed the basis of this attempt, have been brilliantly analyzed by the historian of biology William Coleman. Coleman shows how this attempt and these attitudes emerged in the context of Bateson's opposition to the theory of chromosomes, which had been developed by 1925. The theory held, and still holds, that all hereditary phenomena can be traced to a material particle, known as the gene, which is lodged in the chromosome. This atomistic, Newtonian approach sees the gene as the one hereditary element that is stable, persisting through all change. To Bateson such an approach started at the wrong end of the problem. What persisted, as both Samuel Butler and Bateson's next-door neighbor Alfred North Whitehead had told him, was not matter but *form;* what Gregory would later term "Mind." He thus undertook to uncover the pattern and process of evolution by an analysis of heredity and variation, and to do this focused not on regularities, but on deviations from the norm. "Treasure your exceptions," he once remarked to a fledgling scientist; and the elucidation of "normal" anatomy through the study of nature's anomalies became central to his approach. One examined deviations, or morphological disruptions, to find out how the organism in question adapted, how it managed *not* to go to pieces. (Years later, Gregory Bateson would arrive at his own formulations of typical human interaction by studying alcoholics and schizophrenics.) Thus, in his *Materials for the Study of Variation* (1894), a guidebook to animal teratology, Bateson stated that the goal was to ascertain the laws governing form.[5] The origin of variation, he claimed, had to be sought in the living thing itself, not, as Darwin had held, in the environment. Although he was not a Lamarckian, William Bateson, like the early alchemical Newton, saw the principle of transformation as an internal one. To locate the origin of variation in the gene, however, and then to combine this with a theory of fortuitous variation, was to make a late-Newtonian error: to hold that order could somehow emerge from the random collisions of material particles. Newton's later doctrine of change by way of the rearrangement of impenetrable corpuscles was to Bateson an anathema, a nonexplanation.

For Bateson, then, it was not the gene, but the pattern or form of an organism which was the crucial element in heredity; and if so, then symmetry must be the key to the lock. The basic facts of his study came from examining segmentation, such as that which occurs in the earthworm. Biologists call this phenomenon "meristic differentiation," the repetition of parts along the axis of an animal.

This axial symmetry can be distinguished from the type of radial symmetry displayed by starfish or jellyfish. Both types of symmetry show the continuity of cell generations and behavior we call "hereditary." But whereas the segments of the radially symmetric animals are usually all alike, transversely segmented creatures are capable of a dynamic asymmetry between successive segments—"metamerism." In other words, anomalies of merism are the result of a disruption of normal functioning, and this leads to variation; but this process is itself normal. Nonrepetitive segmentation, such as occurs in the development of the lobster's claw, falls into this category. For William Bateson the study of metamerism opened the door to a concrete demonstration of the primacy of form over matter, and enabled a systemic understanding of heredity and variation. As such, his work constituted a first step in developing an alternative to chromosome theory. He eventually came to argue that what was transmitted in heredity was not an objective substance, but the power or faculty of being able to *reproduce* a substance: tendency, disposition, was what was passed on.[6]

Bateson, however, did take one idea from Victorian physics, which was having its own struggles trying to reconcile matter and force. A number of physicists, including Maxwell, had suggested that for heuristic purposes only, the atom be viewed not as a Newtonian billiard ball, but as a smoke ring, or a vortex. The advantages were obvious. The so-called vortex atom made possible an explanation of the universe which was not completely deterministic. The image embodied the unification of matter and force, as Sir Joseph Larmor, its leading exponent, once declared, and it enabled one to talk of force and change without relying totally on Newtonian rearrangement. Like a smoke ring, the vortex atom was seen as being able to twist and divide, producing new loops; and although Bateson did not discuss the vortex atom explicitly, he emphasized spontaneous division as the key characteristic of living matter. His own notion of living matter, derived partly from ideas already current in Cambridge zoological circles, held that an organism was a "vortex of life." In 1907 he wrote that animals and plants were not simply matter, but systems through which matter was passing. Consciousness apart, said William Bateson, any entity that, like a smoke ring, could spontaneously divide had to be regarded as a living entity. There is no vitalism in his work, no assumption of "God" or an *élan vital*. But his is a type of explanation which has little in common with traditional physics, and in fact much more in common with alchemy. In both—as in what would later become information theory—nature is first and foremost "a perpetual circulatory worker."[7]

The image of the vortex—what would later be called, in cybernetic terminology, the concept of circuitry—was, like the argument of the primacy of form over matter, essential to Bateson's repudiation of chromosome theory. If an organism is an integral whole, a system rather than a mere assemblage of "characters," variation is a phenomenon that has serious consequences, for it must precipitate a coordinate change throughout the entire organism. In the nineteenth century, the French physiologist Claude Bernard had spoken of the *milieu intérieur* (internal environment) of an organism—an environment that Walter Cannon, in *The Wisdom of the Body* (1932), saw as being maintained by a process he called "homeostasis." This notion was William Bateson's central holistic principle. He wrote to his sister Anna in 1888:

> I believe now that it is an axiomatic truth that no variation, however small, can occur in any part without other variation occurring in correlation to it in all other parts; or, rather, that no system, in which a variation of one part had occurred without such correlated variation in all other parts, could continue to be a system.

Initial variation thus acts as an environmental change, setting off a chain reaction throughout the "circuit" or "vortex." Some time must elapse before the organism is once again a system. As Gregory Bateson would argue years later, any system, whether a society, culture, organism, or ecosystem, which manages to maintain itself is rational from its own point of view; even insanity obeys a "logic" of self-preservation. As the years went by, William Bateson became increasingly convinced that the interrelations of the parts of a system were subject to geometric control just as concentric waves in a pool, and that the key to the laws of form involved finding the "accomodatory mechanism," or homeostatic principle. Furthermore, he guessed that this "mechanism," which he believed coordinated the organism as a whole, would be a periodic phenomenon, like a wave. During the mid-1920s, the father began to draw the son into his research. They coauthored an article in which this "undulatory hypothesis" was extended to the study of partridges in an attempt to explain how rhythmical banding develops and spreads over the organism, even down to the tips of the feathers. The "analogy with the propagation of wave-motion must, in part, at least," wrote the authors, "be a true guide."[8] Whether this hypothesis is valid or not, it is clear that the concepts and methodology developed by his father formed the matrix of Gregory's early scientific experience. "I picked up a vague mystical feeling," wrote the latter in 1940,

> that we must look for the same sort of processes in all fields of natural phenomena—that we might expect to find the same sort of laws at work in the structure of a crystal as in the structure of society, or that the segmentation of an earthworm might really be comparable to the process by which basalt pillars are formed.

Above all, it was William Bateson's attitude toward reason itself which shaped so much of Gregory's scientific and emotional consciousness. Reason, writes Coleman, was for William not the mere Newtonian shuffling of atomic sense impressions but "the intuitive grasp of essential relations." He saw the vortex atom, or any such scientific model, in the same way he saw an oriental print. It had conceptual wholeness. It inspired the imagination to an understanding not attainable by rational calculation. William Bateson saw this sort of intuitive insight as evidence for the view that there was a limit to the truth of any scientific explanation, and that there was a deeper level of reality (Mind) which lay beyond its reach. This notion of necessary epistemological incompleteness, that the Mind can never know itself, is perhaps the crux of Gregory Bateson's whole metaphysics. And if this is the rock on which modern science has finally foundered, it has also proven to be, in Gregory Bateson's hands, the foundation on which a new science might be built.[9]

To turn, then, to Gregory's work, we can summarize his intellectual development as follows. In the 1920s he studied biology and anthropology, roughly following in his father's footsteps at Cambridge. The 1930s were devoted to anthropological fieldwork, first among the Iatmul people of New Guinea, which resulted in the publication of *Naven* (1936), then among the Balinese, where he collaborated with his then wife, Margaret Mead. Bateson served with the American Office of Strategic Services during the War, and then took part in the postwar Macy Conferences at which modern cybernetic theory was formulated. Soon after he coauthored *Communication: The Social Matrix of Psychiatry* (1951) with psychiatrist

Jurgen Ruesch, and spent roughly the next decade as an ethnologist at the VA Hospital in Palo Alto, California. It was here that he had an opportunity to work with alcoholics and schizophrenics, applying the concepts of cybernetic theory to these "diseases" and generating a novel approach to both of them. This work, as well as his work on interspecies communication during the 1960s, eventually enabled him to elaborate a new theory of learning. Finally, the 1970s were characterized primarily by the attempt to integrate the insights from his previous investigations with a revision of Darwinian theory, a new approach to the problem of evolution resulting in the publication of *Mind and Nature: A Necessary Unity* (1979). With this work Bateson had come full circle, returning to his original interest in biology after having completed one of the most creative intellectual journeys ever undertaken by a single individual. For the purposes of exposition, I shall devote the present chapter to the work in anthropology, ethnology, learning theory, and abnormal psychology, deal with Batesonian epistemology and its ethical implications in Chapter 8, and devote part of Chapter 9 to a critique of Batesonian holism as a future metaphysics.[10]

As Bateson explains, certain biological analogies he learned in the 1920s, and his father's approach to the natural world, led to his study of the Iatmul people of New Guinea. Bateson's investigation focused on the transvestite ceremony known as "naven," but the nature of the ceremony itself proved to be much less important than the fact that the investigation uncovered, in Bateson's eyes, the nature of scientific explanation itself, and ended in the formulation of a model that might explain the essential character of all mental interaction. Since this model and the methodology that generated it contain the seeds of many of Bateson's later theories of social and natural phenomena, it is important to examine his investigation of naven in some detail.[11]

Naven is a ritual performed by the Iatmul in which the men dress like women and vice versa, and then act out certain roles normally associated with the opposite sex. The occasions for naven are the achievements of the *laua*, or sister's child, and the celebration itself is the responsibility of the *wau*, or mother's brother. Hence the essential relationship is between uncle and niece or nephew, but naven is in fact performed by "classificatory" *waus*, not by the actual maternal uncle. "Classificatory" *waus* are relatives related to the *laua* in a matrilinear way, for example, the great-uncle or male

202

relatives who are in a type of brother-in-law relationship to the father of the *laua*.

There is a whole list of standard cultural acts that call for naven, acts which are most important when performed by a boy or girl for the first time. These include (for a boy) the killing of an enemy or foreigner; the killing of certain animals, or the planting of certain plants; using certain types of tools or musical instruments; traveling to another village and returning; marriage; possession by a shamanic spirit, and so on. For a girl the list includes catching fish, cooking sago, or bearing a child, among other instances.

In the ceremony itself, the classificatory *waus* put on bedraggled female costumes, take the name of "mother," and then go searching for their "child," the *laua*. The ritual pantomime might consist of dressing and acting like decrepit widows, and deliberately stumbling about, while the children of the village follow with peals of laughter. When women play a part, the (classificatory) aunts may beat their nephew or niece when his or her achievements are being celebrated. Unlike the men, the women do not dress in filthy garments, but put on the most fashionable male attire. They may paint their faces white with sulfur—the privilege of men who have committed homicide—and carry male ornaments. They are referred to by male family terminology (father, elder brother, etc.), and affect the bravado commonly associated with male behavior among the Iatmul, while the men act in a self-humiliating manner. The ceremony may also include a pantomime reversal of overt sexual activity. Bateson observed one ceremony in which the *mbora*, or *wau's* wife, dressed as a male and simulated the actions of copulation with her husband, taking the male-superior role. Sometimes the *wau* will pantomime giving birth to the *laua*.

From a Western point of view the whole ceremony, with its deliberate confusion of sexual roles and attire, seems totally incomprehensible. What could the Iatmul possibly think they are doing? In trying to answer this, Bateson followed his hunch that the difference between the radial and transverse segmentation of the zoological world had a social analogue. It turned out that the larger Iatmul villages were unstable, always on the point of fissioning along patrilineal lines: father broke away and took his son with him. Unlike the Western situation in which the break is essentially heretical—an ideological difference—the Iatmul situation is *schismatic*. The breakaway group forms another colony, but with the same set of norms as the parent community. The Western model of

203

heresy is similar to metamerism or dynamic asymmetry, whereas the Iatmul model is analogous to radial segmentation, in which the successive units are repetitive.

The problem of social fission, said Bateson, becomes clearer when we realize that the analogy can be stretched to a comparison of how social control is exerted. The mind's eye might conceive of a radially symmetrical animal as being centrifugal, without any controlling center, since the emphasis in the pattern seems to be in the surrounding segments. The Iatmul are similarly centrifugal, because they have no law, no central established authority that imposes sanctions in the name of the whole community. Offences always take place between two "segments," and social sanctions are "lateral" as well. Western society, on the other hand, emphasizes the state versus the citizen. If I rob my neighbor he may be angry, but it is "the Law" that goes after me and takes action against me. If he should attempt a lateral sanction and decide to take the law into his own hands, he might find himself in as much trouble as I am. Because of this high degree of centralization, Western societies can accomodate a new group with new norms only if it is relatively unobtrusive about its existence. Should it advertise its difference from the center, or assault it, that center will launch a determined counterattack. Iatmul society has no such center, and no such rigidly defined norms. Norms for the Iatmul are seen as conventions to be broken—if one wields sufficient personal power. And since male charisma, so essential to Iatmul sexual ethos, is very much admired, the communities are always on the verge of fissioning along patrilineal lines.

It is thus clear that, in social terms, the naven ceremony makes perfect sense. If schisms occur patrilineally, anything that strengthens affinal ties (those that result from marriage) reduces the chances of schismatic break. The affinal links are the weak points in the whole Iatmul social organization, and thus the naven ceremonies, which reinforce and even exaggerate these links, serve to shore up community integration. In fact, without the naven ceremonies, Iatmul villages could not be as large as they are.

Bateson's explanation of the social meaning of naven is brilliant, but the real inspiration here lies in the fact that he never took his own explanation seriously. Given that naven serves the function indicated, can anyone really believe that the powerful emotive energy evident in the ceremonies is explicable in sociological terms? Would anyone seriously wish to assert that transvestism

204

and ritual copulation are performed for the express purpose of preventing social fission? Bateson was aware that this type of explanation lacked an understanding of the motives of its participants, and he realized that the clue to such motives lay in the "ethos" of the culture, its overall emotional climate. If one wanted the ethos, which was as much a matter of value as it was of fact, one would have to formulate a new definition of scientific methodology. The strictly functional/analytical approach is correct in some rational or pragmatic sense, but it misses the whole point. As his father had once written, it was easier for a scientist to solve a difficulty than to feel it. Gregory had found a situation in which feeling and solving were two sides of the same coin.

What to use as a model? As much as Bateson was impressed by the analytical work of famous contemporaries such as Bronislaw Malinowski and Edward Evans-Pritchard, his real mentor was the great anthropologist Ruth Benedict, whose concept of "configuration" pretty much corresponded to what he was to call the sum of ethos and "eidos." Ethos was the general emotional tone of a culture, eidos the underlying cognitive ("logical") system that a culture possessed. The "concepts," wrote Bateson,

> are in all cases based upon an holistic rather than a crudely analytic study of the culture. The thesis is that when a culture is considered as a whole certain emphases emerge built up from the juxtaposition of the diverse traits of which the culture is composed.[12]

Hence the abstract property, the "feel" or ethos, arose from the *arrangement* of concrete elements. It could not be located in the same way that the elements could, for it was of what he would later call a higher "logical type" than they. A different juxtaposition would necessarily mean a different culture, even if all the elements were identical. In this way, said Bateson, we can state that a culture affects the psychology of its individuals without also stating that a Hegelian *Zeitgeist*, or Jungian "group mind," is somehow at work. Following the example of Ruth Benedict, he continued,

> I shall speak of culture as *standardising* the psychology of the individuals. This indeed is probably one of the fundamental axioms of the holistic approach in all the sciences: that the object studied—be it an animal, a plant or a community—is composed of units, whose properties are in some way *standardised* by their position in the whole organisation. . . . Culture will affect their scale of values. It will affect

the manner in which their instincts are organised into sentiments to respond differently to the various stimuli of life.

As Bateson admitted, the method was deliberately circular: you determined the system of sentiments normal to the culture (the ethos), then invoked it as an explanation for institutions and behavior. Such circularity, he held, should not be a problem, because it could be avoided only by taking a functional, or sociological, view of the system, and this told you nothing about the *motives* of individuals. If you wished to know motives, you had to put yourself *inside* the system, and to do this was inevitably to plunge into circularity. There was nothing mysterious about this situation, as even Gödel's theorem showed. Our behavior was no less real for being self-validating.

What, then, would constitute an adequate analysis of Iatmul ethos, and what might this analysis tell us about their reasons for performing naven? Much of the ethos of both Western and Asian societies arises out of social differentiation, especially that between classes or castes. How one behaves in the presence of another, the emotional tone one adopts, is at least partly conditioned by relative social position, the importance of which varies from one society to the next. In Iatmul culture, on the other hand, there are no social classes and differentiation occurs according to sexuality. Hence, Bateson's chapters on Iatmul ethos are necessarily discussions of *sexual* ethos.[13] He asks: How do the men act with each other, how do the women act with each other, and how do the two sexes act in mixed company?

The dominant characteristic of male behavior in public situations, whether in mixed or all-male company, is pride. For Iatmul men life is virtually a theatrical performance, and activities performed in the ceremonial house incline toward the spectacular and the violent. The house is both a place of ritual and a place for debating and brawling, but it is this latter aspect that largely prevails. As Bateson notes, in the Iatmul mind the ceremonial house is "hot," pervaded by "a mixture of pride and histrionic self-consciousness." Entry into the house is marked by some bit of theater: the man coming into public view will swagger or react with buffoonery. Just as the society has no law or central authority, it has no hierarchy of power, no chieftains. What it has, instead, is a "continual emphasis on self-assertion." Standing is attained by way of the achievements of war, shamanism, esoteric knowledge, and also by playing up to the public.

This behavior is especially marked during public debates that seek to resolve some point of conflict. "The speakers," writes Bateson, "work themselves up to a high pitch of superficial excitement, all the time tempering their violence with histrionic gesture and alternating in their tone between harshness and buffoonery." A speaker might threaten to rape members of the opposition, for example, and pantomime his threat with an obscene dance. When some speaker finally manages an insult too great for the opposition to tolerate (usually by making fun of their totemic ancestors), a brawl erupts which may lead to heavy injuries, and eventually to feuds that involve killing by sorcery.

Although the ceremonial house is for men only, the reaction of the village women is never far from the men's minds. Activities in the house are a preparation for public ceremonies, in which the men perform before the women in full regalia. Initiations, which take place in the house, are deliberately staged so that parts of the ceremony are visible to the women who are nearby and outside, and who thus form an audience. The women also hear the sounds of the secret tribal instruments, and "the men who are producing these sounds are exceedingly conscious of that unseen audience of women." The whole culture, says Bateson, "is moulded by the continual emphasis upon the spectacular, and by the pride of the male ethos."

As might be expected, the ethos of Iatmul women is quite the reverse, though there are instances of remarkable female assertiveness which are regarded as admirable by the Iatmul people. But for the most part, if the lives of the men are preoccupied with "theater," those of the women are centered around "reality": obtaining and cooking food, keeping house, and rearing children. Such activities are done privately, with no regard to appearances. Female style is unostentatious, sometimes to the point of being taciturn. The general spirit is one of quiet jollity and cooperation, and it is the theatrical behavior of the men which provides most of the drama of a woman's life. When, however, the women are collectively called upon to dance publicly, they exhibit a proud ethos, wearing male ornaments and even moving with a slight swagger. This mild transvestism becomes full-blown during naven.

As Bateson points out, his sketch of the Iatmul sexual ethos is drawn from a European point of view. Male Iatmul behavior is histrionic to us but not to the Iatmul, who find it quite normal. If we put ourselves within the culture, we learn that the women find the men strong and assertive, whereas the men find female behavior weak, sentimental, and even shameful. Taking the female position in sexual intercourse is seen as degrading, and for this reason the role reversal during naven's simulated copulation is almost shocking. We see, then, that "each sex has its own *consistent* ethos which contrasts with that of the opposite sex." Naven is remarkable for its reversal of these two very rigid cultural styles.

We are finally in a position to understand why naven is performed. The immediate motivation is tradition: a child has achieved something noteworthy and its relatives must therefore publicly express their joy. In this sense naven is no more esoteric than a bar mitzvah. What we are really asking is why the celebration takes its particular form, for obviously, the Iatmul might simply celebrate with a feast. The sexual ethos described above provides the answer. The men are accustomed to a theatrical display of emotion, not the genuine expression of it. Women, on the other hand, are allowed to express real joy in the achievements of others, but are rarely involved in spectacular public behavior. The child's achievement, however, forces the Iatmul to enact a celebration that cuts across this rigid sexual categorization, violating the norms of both sexes. The men can identify with public display, but not with expressions of joy. The women can express joy, but to make a public display is to violate their norm. The result is acute embarrassment for both sexes, and it is this embarrassment that pushes the situation towards transvestism.

Bateson's point of comparison here is the fashionable English horsewoman who wears decidedly masculine clothing when she rides. Horse riding, compared to the more typically approved "female" activities of British culture, has a definite masculine flavor, generating as it does a powerful sense of physical mastery. In Britain no less than in New Guinea men and women are socialized along very different lines. When she rides a horse the British woman is placed in a situation somewhat unusual for females, but typical for men; hence a masculine costume appropriate to an "abnormal" situation. Similarly, the Iatmul woman engaging in a public display is doing a man's "thing," but wearing a man's costume takes the edge off the resulting embarrassment. Wearing such a costume says, in effect, "It's really OK, I'm a man right now." As for the man, he wears filthy garments and acts in an ineffectual way because his ethos has taught him to regard female behavior as weak or dispicable. This "exchange" behavior is so emotionally charged that at its peak the *wau* may simulate giving birth, while the *mbora* (*wau*'s wife) may jump on the *wau* and humiliate her husband by taking the active role in ritual copulation.

Although he did not stress it at this point, Bateson did note one other psychological motive for naven besides the mitigation of embarrassment. Most cultures possess an aggressive male ethos of performance (the injunction to "Be a man!"); but in Iatmul culture, Bateson observed, this pressure may be an above-average burden on the emotions. As Jung noted, all personalities have both feminine and masculine components, and it is thus possible that the Iatmul sexual ethos is suffocating even to them. That a man must never express joy for another's achievements, or be passive in sex, and that a woman must never be ostentatious, or the sexual aggressor, probably generates enormous psychological tensions. Clearly, these tensions are a source of energy in the naven ceremony, which affords some relief by allowing each sex to "be" the other for a short time and act out the severely repressed parts of its personality. The very frequency of the naven ceremonial, which is performed upon the slightest excuse, further corroborates the argument that it is a counterbalance to a burdensome sexual ethos. As the Iatmul themselves would say, theirs is a "hot" society, generating powerful tensions that are frequently and dramatically relieved.

Bateson's reflections on the nature of these social and psychological tensions led to the formulation of his greatest anthropological concept, that of schismogenesis. Once again he searched for a social analogue to the biological distinction, emphasized by his father, between radial and meristic differentiation. Relations between Iatmul men built to a climax along symmetrical lines. In the ceremonial house, ridicule was met with ridicule, irony with irony, and boasting with boasting until some remark finally precipitated a brawl. Male-female relations, however, followed a very different pattern. Although we have spoken of a male ethos and a female one, they are hardly independent of each other. The men are theatrical *because* the women admire the show; the women are passive (for the most part) *because* the men are histrionic; and it is likely that the behavior of each invokes increasingly exaggerated reciprocal responses. Thus this form of schismogenesis, which Bateson calls "complementary" in contrast to the symmetrical schismogenesis of male-male relations, also escalates over time and builds to a climax, and we might reasonably wonder why Iatmul society does not simply explode from both types of schismogenesis. Indeed, at least in the case of the symmetrical rivalry, it does, and it is the naven ceremonial that keeps Iatmul society from falling apart completely. Although debates do become brawls, and brawls become long-standing feuds, the practice of naven strengthens affinal links and thus softens the harshness of clan opposition.

One sees here a mixture of the two types of schismogenesis. Relations between *wau* and *laua* are complementary, while the link between brothers-in-law is symmetrical. The *wau-laua* relationship thus acts as a brake on symmetrical schismogenesis. In naven, the *wau* insists on the complementary aspects of his relationship with his *laua* at the expense of the symmetrical aspects of the family setup. He acts as "mother" or "wife" to the *laua*, thus denying his real position as a (classificatory) brother-in-law, which is the symmetrical aspect of the relationship. Naven also prevents a cultural breakdown along sexual lines by allowing men and women to "become" each other, even to the point of switching roles in simulated intercourse, thereby releasing the tension accumulated by progressive personality distortion. Naven thus defuses the climacteric that builds in both symmetrical and complementary schismogenesis; and once the ritual drama has ended, the whole process is ready to begin anew.

In general, Bateson defined schismogenesis as "a process of differentiation in the norms of individual behavior resulting from cumulative interaction between individuals." But he soon realized that the concept was more broadly applicable, for cumulative, or "progressive," behavior seemed to be inherent in numerous types of human social and psychological organization. Bateson did not use "progressive" in its familiar Western sense, which has a melioristic connotation, but instead used the term to describe any type of behavior which built to a climax. In progressive change of this sort, the absence of stabilizing elements usually means that the process will end in explosion or deterioration (a "runaway" situation). To take the most general case, consider two social groups, which we shall label "A" and "B." (These could be men and women, parents and children, two nations, two political factions, etc.) Symmetrical schismogenesis occurs when the two groups get into a relationship that resembles the rivalry at an auction. The two behaviors are identical, with each group attempting to do the opposition one better: "Well, top *this*, then." This sort of rivalry can be seen at work in situations of culture contact, interpersonal competition, and in the whole arena of politics, as the tiresome game of Pentagon-Kremlin arms race clearly shows.

In the case of complementary schismogenesis, the rivalry is reciprocal; the aggressive behavior of A, let us say, provokes submissive behavior on the part of B, encouraging more aggression from A in what becomes an escalating spiral. The classic example is perhaps the traditional marriage, in which the pattern of dominant husband/submissive wife is initially satisfactory to both parties involved. Over time, however, the roles distort one another. The wife's submission provokes the husband's assertion, in turn encouraging her submission, and so on. No one is by nature completely assertive or submissive, but the dynamics of the relationship increasingly repress one side of each partner's personality, until each recognizes the stunted aspect of his or her own personality overdeveloped in the personality of the other. Ultimately each becomes unable to see the other's viewpoint. They have lost all interest in making the relationship work, while reciprocal tensions continue to accumulate. Finally, the husband may be goaded into totally despotic behavior in an attempt to provoke a counterreaction, and the wife may decide to blow her brains out—or his. More typically, she will leave the marriage. This single example illuminates the mechanism of a number of other types of interpersonal or political situations. Complementary schismogenesis can be seen at work in certain cases of culture contact, in numerous types of group behavior (e.g., the reinforcing of one member's "deviant" pattern by the actions of the other members), and in situations such as class conflict or racial oppression.

As in the case of the Iatmul, we must ask why the whole world is not exploding; and again, we are forced to reply: it *is*. Nevertheless, as Bateson recognized, things do not always escalate to breakdown. Some marriages stabilize, though few are happy. The Pentagon and Kremlin may do us in yet, but have managed to avoid Armageddon thus far. Class rivalries are often bitter, but as Marxists have found, industrial societies are not fertile ground for proletarian revolution. Trying to explain why, Bateson theorized that, as in the Iatmul situation, schismogenic tensions were being eased by admixtures. Medieval principalities sometimes had one day a year in which serfs became kings and the king a subject—a single brief role reversal that was often enough to keep the whole system going. The traditional marriage has been feasible up to recent times because the wife could at least be mistress of the kitchen, even if subservient everywhere else. Internal rivalries tear at industrial societies between wars, only to be resolved at a stroke by the appearance of a common enemy, which switches the internal symmetrical tensions into a complementary mode and provides a target on which to focus symmetrical schismogenesis. (Labor and management, for example, now share complementary roles in the effort to defeat a common enemy.)

The concept of schismogenesis also enabled Bateson to reply to the most trenchant criticisms of Ruth Benedict's type of anthropology. Of what real value were the concepts of "configuration," "standardization," and "modal personality," it was asked, if it were obvious that any single society possessed a greater divergence of social types *within* itself than existed between it and other societies? How, for example, would you explain the deviant personality, the individual who has clearly escaped the pressures of his or her context?[14]

As early as 1942, Bateson pointed out that both individuals and societies are organized entities. In the Iatmul researches, it had not been enough to say that the character structure of one sex was very different from the character structure of the opposite sex. The point was that the ethos of one *cogged into* the ethos of the other; that the behavior of each promoted the habits of the other. *All social, personal, and biological life has its own "grammar," or code.* You can react *against* your particular code; but you can hardly behave in a way that is totally irrelevant to it. Furthermore, these patterns tend to be bipolar. If you are trained in one half of such a pattern, it is a fair guess that the seeds of the other half are sown somewhere in your personality. Thus, argued Bateson, it is not that husband and wife are trained, respectively, in dominance and submission. Dominance and submission are integrally (dialectically, alchemically) related; there is no pure dominance or pure submission. The couple was instead trained in dominance/submission, as a total pattern, and given enough dominance from the husband, the wife may assert her repressed dominance in the form of homicide. The fact that the *mbora* can vigorously ape the male sexual role in naven suggests that submission was not all her society taught her. Hence, Bateson concluded that when we deal with relatively stable differentiation within a community, we are justified in speaking of a modal, or standardized, personality if we describe it in terms of the motifs of relationships familiar to the entire community. The deviant personality has *not* escaped the pressures of its community, for its deviance is a reaction *to* those motifs. The deviate's behavior may not follow social norms, but it is acquired with respect to those norms, and even if the behavior is the opposite of those norms, it

212

still retains its relevance to them. The relationship of agent/tool, for example, is absent in New Guinea; Iatmul deviates do not behave in these ways. Or, to take a more famous example from Bateson's work, insanity is just such a reaction to cultural norms. Rather than being a "disease" that descends on the victim from out of the blue, it is a patterned, "logical" response that meshes quite efficiently with the surrounding family structure.[15]

Is schismogenesis truly inherent in human behavior? It is a compelling thesis, yet one that was completely disproved by Bateson's next anthropological investigation, that of Balinese society.[16] Without going into too much detail here, it is important to note that Bateson found the nondialectical situation of the Balinese totally unprecedented. Their culture was not, he realized, susceptible to any type of Hegelian or Marxist analysis. Balinese music and art are characterized by balance, not by tension and resolution, as in the West; and indeed, balance seems to be a metaphor that extends to every phase of Balinese life. The emphasis is on present enjoyment; the Balinese have no concept of future reward, and things are done in and for themselves. Life itself is seen as a work of art. The best metaphor for the Balinese way of life might be a tightrope walker constantly adjusting his balancing pole so as to turn out a graceful and pleasurable performance.

Competition and rivalry are thus absent in Bali. Should a quarrel arise between two members of the society, they will go to a local official and register the fact that they have a quarrel. There is no attempt at reconciliation and, in effect, they have drawn up a contract of enmity. Still, the two enemies are able to recognize their relationship as it is, to accept its existence at that particular plateau, and as a result, climactic interaction is obviated. The Balinese, like the Iatmul, recognize no central authority, but unlike the Iatmul they do not regard offences as personal. If, says Bateson, a caste-less person fails to address a prince formally, the prince sees not a personal insult but an offence against the natural order of the universe, a violation of postural balance. In everything they do, *optimization* is the issue, not maximization. Balinese economics, for example, cannot be described in terms of a profit motive, nor can the Balinese social structure be seen as a collection of individuals or groups vying for status or prestige.

The Balinese apparently achieve this balance through their child-rearing practices, teasing their offspring into cumulative interaction and then deliberately losing interest just shy of the point of climax. In most cultures this technique would produce psychotic individuals, but in Bali the totality of the pattern reinforces the practices and produces adults who distrust cumulative involvement. Still, we must resist our Western assumption that life in Bali must be one tedious attempt to preserve the status quo. Like the Iatmul, we are trapped in the notion that schismogenic situations, which are in fact profoundly neurotic, are exciting, and that anything else must be dull. In one of his best passages in *One-Dimensional Man*, Herbert Marcuse correctly characterized the apparent dynamism of advanced industrial culture as fraudulent: "Underneath its obvious dynamics this society is a thoroughly static system of life: self-propelling in its oppressive productivity and in its beneficial coordination."[17] Factory life, consumer life, business life, executive life—all of these are, from the inside, boring, repetitive, and characterized by an absence of any real adventure or exploration.

The situation in Bali is just the reverse. It looks like a "cool" society, but is in fact very active. The Balinese, says Bateson, extend attitudes based on body balance to human relations. They generalize the idea that motion is essential to any type of balance. Their society is a very complex and busy one, but not in our sense, for theirs is steady state maintained by continual nonprogressive change. In his essay "Style, Grace and Information in Primitive Art," Bateson analyzed a Balinese painting, showing that it had as its message the idea that "to choose either turbulence or serenity as a human purpose would be a vulgar error." The Balinese recognize that these poles are mutually dependent in art, sex, society, and death, but they have come to terms with this reality by means of a nonschismogenic solution. Although Bateson never believed in "primitive" solutions for the West, Bali served as an important model for him, acting as a kind of mirror in which the folly of most human interaction was sharply revealed and contrasted.

Schismogenesis, then, is *learned*; it is as much an acquired habit as is the nonschismogenic behavior characteristic of Bali. Yet it seems so fundamental that we are forced to ask what learning itself consists of, if it so inseparably links cognition and emotion (eidos and ethos). What does it mean to learn something, to "know" something? After the Macy Conferences on cybernetics, Bateson made this question the subject of his next major investigation.

Bateson began his study of learning theory with an ostensibly nonsensical question: Is there such a thing as a "true error"? More

214

broadly, is there such a thing as a true ideology? Ideologies are cultural things learned in cultural contexts, but they usually work for the cultures that believe in them. The Balinese believe certain things about the world that to us, or to the Iatmul, seem almost inconceivable. Bateson had regarded cumulative interaction as an inherent trait, but Bali showed him that an entire nation could learn to do something quite different. Furthermore, Balinese society was far more stable than Iatmul or Western European society, and thus in some sense its "crazy" premises had to be more true. Put in this way, the crucial question became, How are ideologies (perceptions, world views, "realities") and emotive patterns (dominance/submission, succor/dependency) formed within the mind of an individual or his society? In response, Bateson followed Benedict's notion of configuration and returned to the concept of the "grammar," or code. Individuals and societies are organized entities; they are "coded" in a certain way that is coherent, that makes sense in both emotional and cognitive terms. Since it was this process of coding which rendered them stable (so long as the code continued to work), it was essential to explain that process more fully.[18]

Down to some point in the mid-1960s, learning theory was dominated by the behavioral model, which is most commonly associated with J. B. Watson and B. F. Skinner. The real grandfather of such work was Ivan Pavlov, who had managed to immortalize a dog by getting it to salivate when he rang a bell. What Pavlov did was to set up a context of association. Repeatedly, bell-ringing was followed by food, until the sound of the bell alone was enough to trigger the animal's entire gastronomic response. In one of Skinner's experiments, a rat learned to press a bar and thereby release a pellet of food. Skinner's rat had to contend with a set of rules different from those that confronted Pavlov's dog, but again a context of (causal) association was central: event occurs, food appears. Furthermore, all of these experiments involved a progressively faster rate of learning on the part of animals. Dog and rat quickly caught on to the rules of the game. After a number of trials, the dog did not need meat to salivate; he had learned what the bell meant. Similarly, the rat discovered that the food pellet was no accident, and began to spend much of its time pressing the bar.

What is going on in such experiments? What do the terms "learn" and "discover" mean, as I have just employed them? Bateson uses the term "proto-learning" to characterize the simple solu-

215

tion of a problem. Bell rings, or bar is presented. The Pavlovian situation requires a passive response, the Skinnerian situation a more active one, but there is still a problem to be solved in each case: what does this phenomenon require of me (dog, rat), and what does it lead to? Solving such a specific problem is proto-learning, or Learning I. "Deutero-learning," or Learning II, Bateson defines as a "progressive change in [the] rate of proto-learning." In Learning II the subject discovers the nature of the context itself, that is, he not only solves the problems that confront him, but becomes more skilled in solving problems in general. He acquires the habit of expecting the continuity of a particular sequence or context, and in so doing, "learns to learn." There are, furthermore, four contexts of positive learning, as opposed to negative learning, in which the subject learns *not* to do something. There are the two already described, Pavlovian contexts and those of instrumental reward; and there are also contexts of instrumental avoidance (e.g., rat gets an electric shock if it doesn't press the bar within a certain time interval) and of serial and rote learning (e.g., word B is always to be uttered after word A). So proto-learning is the solution of a problem within such contexts, and deutero-learning is figuring out what the context itself is—learning the rules of the game.

Character and "reality" have their origins in the process of Learning II; indeed, character and reality prove to be inseparable. A person trained by a Pavlovian experimenter would have a fatalistic view of life. He would believe that nothing could affect his state, and for such a person reality might well consist of deciphering omens. A Skinnerian-trained individual would be more active in dealing with his or her world, but no less rigid in his or her view of reality. Western cultures, notes Bateson, operate in terms of a mixture of instrumental reward and avoidance. Its citizens deutero-learn the art of manipulating everything around them, and it is difficult for them to believe that reality might be arranged on any other basis. The link between fact and value is (a) that such acquired perceptions are also acquired character traits, and (b) that they are purely articles of faith. In other words, to take (a) first, any bit of learning, especially deutero-learning, is the acquisition of a personality trait, and *what we call "character" (ethos, in Greek) is built on premises acquired in learning contexts.* All adjectives descriptive of character, says Bateson—"dependent," "hostile," "careless," and so on—are descriptions of possible results of Learning II.

216

The Pavlovian-trained person not only sees reality in fatalistic terms; we might also say of him or her, "She is fatalistic," or "He is a passive type." Most of us raised in Western industrial societies have been trained in instrumental patterns, and therefore we do not ordinarily notice these patterns: they constitute our ethos. They are "normal," and thus invisible. In especially egregious cases, we will say, "He's only out for himself"—a character description that is at the same time an epistemology. Dominant, submissive, passive, self-aggrandizing, and exhibitionistic—all are simultaneously character traits and ways of defining reality, and all are (deutero-) learned from early infancy.

The second point, that these "realities" are articles of faith, raises the issue of the "true ideology." If you have been raised with an instrumental view of life, you will relate to your social and natural environment in that way. You will test the environment on that basis to obtain positive reinforcement, and if your premises are not validated, you will probably not abandon your world view, but classify the negative response, or lack of response, as an anomaly. In this way you remove the threat to your view of reality, which is also your character structure. Neither the witch doctor nor the surgeon gives up magic or science when his methods fail, as they often do. Behavior, says Bateson, is controlled by Learning II, and molds the total context to fit in with those expectations. The self-validating character of deutero-learning is so powerful that it is normally ineradicable, usually persisting from cradle to grave. Of course, many individuals go through "conversions" in which they abandon one paradigm for another. But regardless of the paradigm, the person remains in the grip of a deutero-pattern, and goes through life finding "facts" that validate it. In Bateson's view, the only real escape is what he calls Learning III, in which it is is not a matter of one paradigm versus another, but an understanding of the nature of paradigm itself. Such changes involve a profound reorganization of personality—a change in form, not just content—and can occur in true religious conversion, in psychosis, or in psychotherapy. These changes burst open the categories of Learning II itself, with magnificent or hazardous results. (We shall deal with Learning III at greater length below.)

It should be clear, then, that the union of fact and value, which modern science denies in principle, occurs quite naturally in Bateson's analysis of learning. A system of codification, he says, is not very different from a system of values. The network of values partially determines the network of perception. "Man lives by those propositions whose validity is a function of his belief in them," he writes. Or as he says at a later point, "faith is an acceptance of deutero-propositions whose validity is *really* increased by our acceptance of them."

But what *is* character structure? If it was an error to reify ethos in New Guinea, Bateson realized, it was no less fallacious to treat a character trait as a thing. Adjectives descriptive of character are really descriptions of "segments of interchange." They are descriptions of *transactions*, not of entities, and the transactions involved exist between the person and his or her environment. No person is "hostile" or "careless" in a vacuum, despite the contrary contention of Pavlov, Skinner, and the whole behavioral school. Clearly, Learning II is equivalent to the acquisition of apperceptive habits, "apperception" being defined as the mind's perception of itself as a conscious agent. Such habits can be acquired in more than one way, and the behaviorist is wrong to believe that habit is formed only through the repeated experience of a specific kind of learning context. "We are not concerned," writes Bateson,

> with a hypothetical isolated individual in contact with an impersonal events stream, but rather with real individuals who have complex emotional patterns of relationship with other individuals. In such a real world, the individual will be led to acquire or reject apperceptive habits by the very complex phenomena of personal example, tone of voice, hostility, love, etc. Many such habits, too, will be conveyed to him, not through his own naked experience of the stream of events, for no human beings (not even scientists) are naked in this sense. The events stream is mediated to them through language, art, technology and other cultural media which are structured at every point by tramlines of apperceptive habit.[19]

The psychology laboratory is probably the *last* place to learn about learning, just as the physics laboratory is the last place to learn about light and color. Both Skinner and Newton were guilty of narrowing the context to the point that they could have precise control over the trivial. If you wish to find out about learning, contends Bateson, study individuals in their cultural context, and study especially the non-verbal communication that goes on between them. Deutero-learning proceeds largely in terms of what he would later call "analogue," as opposed to "digital" cues. It is in this arena that we shall, he believed, find the source of our character "traits" and our cognitive "realities."

To enlarge on this for a moment, digital knowledge, which expanded rapidly after Gutenberg's time, is verbal-rational and abstract. For example, a word has no particular relationship to what it describes ("cow" is not a big word). Analogue knowledge, on the other hand, is iconic: the information represents that which is being communicated (a loud voice indicates strong emotions). This kind of knowledge is tacit, in Polanyi's sense, and includes poetry, body language, gesture and intonation, dreams, art, and fantasy. Pascal and Descartes had debated this distinction between style and nuance on the one hand, and measurement and geometry on the other. Although at first glance these two forms of knowledge may seem irreconcilable, Bateson chose to believe that Pascal was right when he wrote that the heart had its reasons which reason did not perceive. Perhaps it was time for scientists to start formulating some cardiac algorithms.

Bateson recalls that it was in January 1952, while watching monkeys playing at the Fleishhacker Zoo in San Francisco, that he realized that their play (the monkeys' captivity notwithstanding) could provide a foothold on the whole area of nonverbal communication. The resulting article, "A Theory of Play and Fantasy," argued: (1) that play between mammals dealt with *relata*, rather than manifest content, and in this way was very similar to primary-process material (or dream and fantasy) in its structure; (2) that although it was not familiar to our conscious minds, such material was subject to the analysis of formal logic, specifically the rules of paradox described by Russell and Whitehead in their classic work, *Principia Mathematica* (1910–13), (3) that since humans were mammals, our own learning—and therefore our character and world view—depended on such material; that what we called "personality" and "reality" were formed by a (deutero-) learning process that permeated our environment and taught us, in ways that were subtle but definite, certain allowable patterns that the culture labeled "sane"; and (4) that conversely, insanity (ostensible lack of coherence of personality and world view) probably involved the inability to manipulate the relationship between conscious and unconscious according to the deutero-propositions of a particular cultural context.

The theoretical starting point for Bateson's research here was Russell and Whitehead's "Theory of Logical Types." In itself, the theory simply states that no class of objects, as defined in logic or mathematics, can be a member of itself. Let us, for example, conceive of a class of objects consisting of all of the chairs that currently exist in the world. Anything we customarily term "chair" will be a member of that class. But the class itself is not a chair, any more than a particular chair can be the class of chairs. A chair, and the class of chairs, are two different levels of abstraction (the class being the higher level). This axiom, that there is a discontinuity between a class and its members, seems trivially obvious, until we discover that human and mammalian communication is constantly violating it to generate significant paradoxes.

One of the most famous of these paradoxes is known as "Epimenides' Paradox," or the "Liar's Paradox" (see Chapter 5, note 30). It might be presented as in Figure 14:

```
┌─────────────────────────────────┐
│                                 │
│        All statements           │
│        within this frame        │
│        are untrue.              │
│                                 │
└─────────────────────────────────┘
```

Figure 14. Gregory Bateson's illustration (modified) of Epimenides' Paradox (from "A Theory of Play and Fantasy").

We see the problem at once. If the statement is true, it is false, and if false, it is true. The resolution lies in the Russell-Whitehead axiom. The word "statements" is being used in both the sense of a class (the class of statements) *and* as an item within that class. The class is being forced to be a member of itself, but since this situation is not allowable according to the formal rules of logic and mathematics, a paradox is generated. The statement itself is being taken as a premise for evaluating its own truth or falsehood, and thus two different levels of abstraction, or logical types, are being scrambled.

Now the truth is that neither human nor mammalian communication conforms to the logic of *Principia Mathematica*. In fact, all meaningful communication necessarily involves metacommunication—communication about communication—and is therefore constantly generating paradoxes of the Russellian type. Let us take human communication first. Suppose I announce to you, just as we embark on some particular action or conversation, "This is play." The message I am conveying is, "Do not take the following seriously." What does the phrase really mean? "This is play," says Bateson, can be translated into the statement: "These

actions in which we now engage do not denote what those actions
for which they stand would denote"; or, since "stand" and "denote"
mean the same thing here, the translation can be rendered: "These
actions, in which we now engage, do not denote what would be
denoted by those actions which these actions denote." If I give my
lover a playful nip, the nip denotes a bite, but it does not denote
what would be denoted by an actual bite. It is *not* an act of aggression, and I express this by using the act to comment negatively on
itself. But neither this behavior, nor the statement "This is play,"
are allowable in formal logic. The "translated" sentence is a good
example of the Liar's Paradox: the word "denote" is being used in
two degrees of abstraction, which are incorrectly treated as though
they were on the same logical level (one is allowed to contradict the
other). Both the nip, and the statement "This is play," set up a
frame which is then allowed to comment on its own content.

This discussion returns us to the Fleishhacker Zoo, and the question of what we can learn from monkeys. Metacommunicative
messages are logically inadmissable, because they are frames that
comment on their own content. This point is clear enough in the
case of a verbal statement, such as "This is play," and in fact we are
constantly checking the frame of reference in ordinary discourse:
"What are you really saying?" "Do you mean that?" "You've *got* to
be kidding!" and so on. But, says Bateson, although we are capable, unlike monkeys, of spoken or written metacommunication, we
are like them in the following crucial sense: the vast majority of
metamessages remain implicit. "I love you," I say absentmindedly
to my lover who just walked into the room in search of my atten
tion or affection, while my body language and tone of voice say,
"Leave me alone so I can finish writing my chapter on Bateson."
As for our mammalian cousins, they are limited by the lack of
language such that they can refuse or reject an action, but not
negate or deny it. Two dogs meet, and neither wishes to fight.
They are unable to say, "Let's not fight." Being friendly does not
solve the problem either, because it is a positive statement that
omits any "discussion" of fighting, rather than specifically decid-
ing against it. So the dogs bare their fangs, stage a mock brawl, and
then stop. The message exchanged: the nip is not the bite, or "These
actions, in which we now engage, do not denote . . . , etc." Play
is a phenomenon in which actions of play denote other actions of
not-play; and like dogs or monkeys, we exchange such messages
all the time. In fact, says Bateson, on the human level we have

evolved some very complex games based on a deliberate confusion of map and territory. Catholics say that the wafer is the body of Christ, a sacrament. Protestants say it is *like* the body of Christ, a metaphor. Millions have been killed in war, tortured or burned to death over just this issue, and millions continue to die for this or that flag—bits of cloth which are much more than metaphors in the eyes of the soldiers who march under them.

Here, then, is the similarity between animal communication and primary process. Like dreams and fantasy, play deals (though not exclusively) with *relata* rather than content. The significant message in any dream lies in the relationships between the things in the dream. The image employed in the expression of the relationship is less important than the relationships themselves. Unlike secondary process, primary process cannot comment on itself directly.[20] Map and territory are equated. The frame itself, as Bateson says, becomes part of the premise system; it is metacommunicative. Every fantasy, for example, includes the implied message, "This is not literally true."

Finally, we must ask: So what? So what if most of our communication violates some abstract theory of logic that was formulated in the first decade of the twentieth century? The significance lies in the fact that *it is largely this violation of logic which constitutes most of our deutero-learning;* that we obtain a personality, and a world view, by means of a pervasive system of cultural, metacommunicative messages that can be understood in fairly precise terms; and that in comparison to deliberate, conscious, digital knowledge, this analogue knowledge is incredibly vast. I shall return to this last point, the "principle of incompleteness," in Chapter 8. For now, it is important to understand where Bateson's investigation of learning theory took him. Whereas traditional scientific investigation scrupulously avoids any overlapping of fact and value (this, as we have seen, is the cause of its disembodied quality), Bateson deliberately merged the two, or rather, he did not force the usual artificial separation. As a result, the answer that emerged was both precise and meaningful. A person's "truth" is also his or her "character," and the patterns of formation are to be found in the modalities of nonverbal, or meta, communication.

The work that Bateson is probably best known for, his study of insanity and the formulation of the theory of the double bind as the formal etiology of schizophrenia, is in fact a brilliant elaboration, and verification, of the above theory of learning. As well as any other example we might give, this work illustrates clearly the genesis of world view and personality, and reveals the "cardiac algorithms" that underlie the process. It is a type of proof by counterexample, however, a *reductio ad absurdum,* for madness shows what happens when the ability to metacommunicate is absent, or severely attenuated. What, Bateson asked, was being learned, or rather mis-learned, in the manufacture of madness?

Bateson's exploration of learning theory thus far had led him to the conclusion that the metacommunications system of our culture taught us how to use frames, and that their use defined personality, world view, and social sanity. It was the thesis of one of Bateson's colleagues, the psychiatrist Jay Haley, that the symptoms of madness might be due to an inability to discriminate between logical types. The individual who took metaphor for reality, who insisted, *outside* of church, that the wafer really *was* the body of Christ, was classified as psychotic. In the paper on play and fantasy, Bateson had been content to ask: "Is there any indication that certain forms of psychopathology are specifically characterized by abnormalities in the patient's handling of frames and paradoxes?" The seminal paper on the double bind, written together with Haley, Don Jackson, and John Weakland, appeared the following year, and suggested an affirmative answer.

As far as Haley was concerned, the ability to distinguish between the literal and the metaphorical was the touchstone of sanity. Bateson himself points to the situation in which the schizophrenic patient comes into the hospital canteen, and the woman behind the counter says to him, "What can I do for you?" He does not reply, "I'll have the steak and kidney pie today," but instead he stands there trying to figure out what sort of a message this is. Is she offering to sleep with him? Is she trying to do him in? Is she going to give him a free lunch if he asks for it? The point, says Haley, is that all human messages violate the Theory of Logical types. There is *always* an accompanying metacommunication, usually a nonverbal one, and sanity is the ability to decipher and use this code. Our patient would be correct in concluding that a sexual advance was being made if a certain tone of voice, or body language, had accompanied the woman's question, but he is unable to make such a discrimination, and it is this inability that justifies the label, "insane." Haley gives the example of a schizophrenic man who had been given ground privileges and abused them, escaping from the institution by climbing the fence surrounding it. The police finally

223

found him and brought him back. A few days later the man showed Haley the point in the fence where he went over and said, "There's a stop sign there now." As he spoke, however, there was a twinkle in his eye. Haley suddenly realized that the patient was not being literal. Rather, he had learned to comment on his own messages, and was thus on the road to recovery.[21]

How does a person get to the point of constantly confusing logical types? Bateson believed that we should not look for some childhood trauma, some watershed event, but instead examine what was *regular* in the childhood of the schizophrenic. Somehow, he or she had been trained *not* to metacommunicate, not to comment on the messages of others, and such an inability was so aberrant that it was doubtful that one single incident could have precipitated it. In cybernetic metaphor (see Chapter 8), metacommunication is feedback, and the psychotic is like a self-correcting system that has lost its governor, endlessly spiraling into distortions labeled "catatonia," "hebephrenia," "paranoia," and so on. In fact, these distortions are alternatives to commenting on the messages of others, which for some reason the schizophrenic feels he or she must not do.

What Bateson, Haley et al. did was to investigate the entire family situation, rather than (as is still the norm) the isolated schizophrenic. Bateson and his fellow workers believed that the patient was gripped not by a "disease" mysteriously caused by genes or brain chemistry, but by a *process*, a *pattern*, that had been going on for years. As R. D. Laing, whose own work was based on the "double bind" theory, has shown, the difference between treating a schizophrenic as an "organism" emitting "signs of disease," and as a person engaged in a process, is the difference between night and day. In *The Divided Self*, Laing reproduces the famous account (1905) provided by the German psychiatrist Kraepelin of the latter's presentation of a schizophrenic patient to a lecture room full of students:

The patient I will show you today has almost to be carried into the rooms, as he walks in a straddling fashion on the outside of his feet. On coming in, he throws off his slippers, sings a hymn loudly, and then cries twice (in English), "My father, my real father!" He is eighteen years old, and a pupil of the Oberrealschule (higher-grade modern-side school), tall, and rather strongly built, but with a pale complexion, on which there is very often a transient flush. The patient sits with his eyes shut, and pays no attention to his surroundings. He does not look up even when he is spoken to, but he answers beginning in a low voice, and gradually screaming louder and louder. When asked where he is, he says, "You want to know that too? I tell you who is being measured and is measured and shall be measured. I know all that, and could tell you, but I do not want to." When asked his name, he screams, "What is your name? What does he shut? He shuts his eyes. What does he hear? He does not understand; he understands not. How? Who? Where? When? What does he mean? When I tell him to look he does not look properly. You there, just look! What is it? What is the Matter? Attend; he attends not. I say, what is it, then? Why do you give me no answer? Are you getting impudent again? How can you be so impudent? I'm coming! I'll show you! You don't whore for me. You mustn't be smart either; you're an impudent, lousy fellow, such an impudent, lousy fellow I've never met with. Is he beginning again? You understand nothing at all, nothing at all; nothing at all does he understand. If you follow now, he won't follow, will not follow. Are you getting still more impudent? Are you getting impudent still more? How they attend, they do attend," and so on. At the end, he scolds in quite inarticulate sounds.

Kraepelin added the following notes to this description:

Although [the patient] undoubtedly understood all the questions, *he has not given us a single piece of useful information. His talk was . . . only a series of disconnected sentences having no relation whatever to the general situation.* [Italics Laing's]

Now what is going on here? The sort of "word-salad" reproduced above is very common among schizophrenic patients, and it was Bateson's contention that since the crux of insanity was the inability to metacommunicate, such "word-salad" must contain a comment on the situation, but in a safe, that is, indirect and disguised, form. In fact, unbeknownst to Kraepelin, the patient was parodying the whole interview, and in such a way that allowed him to tell Kraepelin to fuck off: "You want to know that too? I tell you who is being measured and is measured and shall be measured. I know all that, and could tell you, but I do not want to." "This seems," Laing comments, "to be plain enough talk. Presumably he deeply resents this form of interrogation which is being carried out before a lecture-room full of students. He probably does not see what it has to do with the things that must be deeply

225

distressing him." Thus when Kraepelin asks him his name, he replies in a way that comments on Kraepelin's whole approach to him:

> What is your name? What does he shut? He shuts his eyes. . . . Why do you give me no answer? Are you getting impudent again? You don't whore for me [i.e., says Laing, he feels that Kraepelin is objecting because he is not prepared to prostitute himself before the whole classroom of students] . . . such an impudent, shameless, miserable, lousy fellow I've never met with.[22]

From Laing's point of view, Kraepelin is something of a dolt. At another point, Laing relates the story of the patient who was similarly contemptuous of his psychiatrist but was afraid to confront him. Instead, he told him that he heard voices, and when asked what they were saying, looked directly at the doctor and replied: "You are a fool." The psychiatrist busily wrote it down in his note pad.

The question is, why metacommunicate in such an arcane way? Why didn't the boy simply turn to Kraepelin and say, "I object to being treated like a performing bear. Please leave me alone"? Even if Kraepelin had been capable of hearing such a statement, the patient would not have been constitutionally capable of making it, for he had undoubtedly been dealt with by people like Kraepelin all his life. His family situation was probably such as to rule out any overt metacommunication. Hence, word-salad, "validating" Kraepelin's diagnosis. Bateson's hypothesis was that this word-salad was descriptive of an ongoing traumatic situation that involved a tangle in metacommunication, and that this ongoing trauma "must have had *formal* structure in the sense that multiple logical types were played against each other. . . ." A visit to the home of one of Bateson's own patients, for example, revealed that the patient's mother was constantly, and without any apparent awareness, taking the messages received from the people around her (Bateson included) and reclassifying them to mean something else. The patient undoubtedly had to endure such behavior since infancy. But it was he, rather than she, who was judged insane, because it was she, rather than he, who ran the household, and who presumably obtained her husband's support or acquiescence. By the time the son was old enough to say, "That's not what I meant; you are misunderstanding me," he was totally unable to do so. What developed instead was an array of bizarre symptoms.

226

The researches of Bateson and his fellow workers tended to support the general hypothesis that in the psychology of real communications, the Theory of Logical Types (the discontinuity between a class and its members) was constantly being breached. They found that schizophrenia was the result of certain formal patterns of this breaching occurring, in an extreme form, in the communication between mother and child. Of course, metacommunication can always be falsified: the false laugh, the artificial smile. But most typically, as in the above example of mother and son, the falsification was done unconsciously. Like Mrs. Malaprop, the mother was unaware that she was getting everything scrambled, but in this case the consequences were not quite so humorous. At this point in the analysis of schizophrenia, Bateson's theory of deutero-learning became relevant. The son had been deutero-trained into a schizophrenic reality; he had learned to construct reality this way in order to survive. Given this ethos, insanity had become his "character" and world view. But there had to be more to it than that. This was only the beginning of an explanation; what Bateson was seeking was a full-fledged scientific understanding of the phenomenon.

In New Guinea, Bateson had grasped the ethos of the Iatmul, at least in part, through the concept of schismogenesis. Did schizophrenia also have such formal structure, and if so, what was it? What did Learning II consist of, for psychotic individuals? The "road through the mystery of Species," William Bateson had written in 1894, "may be found in the facts of Symmetry."[23] What was the symmetry in this case? What was the underlying pattern, the cardiac algorithm? The schizophrenic child, wrote Bateson et al., lives in a world in which sequences of events are such that unconventional habits of communication are in some sense logical. "The hypothesis which we offer," the authors continued, "is that sequences of this kind in the external experience of the patient are responsible for the inner conflicts of Logical Typing. For such unresolvable sequences of experiences, we use the term 'double bind.'" Bateson identified the ingredients of a double-bind situation as follows:

(1) Two or more persons must be involved, one of whom is forced to play the role of victim.

(2) The double-bind structure goes on repeatedly. It is not a matter of some great traumatic shock, but of a regular and habitual way of experiencing the world.

(3) There is a primary negative injunction, either of the form,

227

"Do not do X, or I will punish you," or "If you do not do X, I will punish you." Again, the punishment is not a key traumatic event, but an ongoing one, such as withdrawal of love or expression of abandonment.

(4) There is a "secondary injunction conflicting with the first at a more abstract level, and like the first enforced by punishments or signals which threaten survival." Here is the confusion of logical types. The secondary injunction is usually (meta)communicated by kinesic signals. The parent, for example, might punish the child, and then display a body language that says "Do not see this as punishment," "Do not see me as the punishing agent," or even, "Do not submit to this." In acute forms of schizophrenia, parents do not have to be present anymore. "The pattern of conflicting injunctions may," says Bateson, "even be taken over by hallucinatory voices."[24]

(5) However, the double bind is not merely a "damned if you do, damned if you don't" situation. In and of itself, a no-win situation cannot drive someone crazy. The crucial element is not being able to leave the field, or point out the contradiction; and children often find themselves in just such a situation. Thus Laing sums up the double-bind predicament as: "Rule A: Don't. Rule A.1: Rule A does not exist. Rule A.2: Do not discuss the existence or nonexistence of Rules A, A.1, or A.2."[25]

What happens to a child caught in such a situation? Clearly, he will have to falsify his own feelings, convince himself that he really doesn't have a case, in order to maintain the relationship with his mother or father. In formal terms, he will have to (deutero-) learn *not* to discriminate between logical types, because it is just such discrimination that will threaten the whole relationship. In other words, (a) he is in an intense relationship, hence feels he must know what messages are being communicated to him; (b) the person doing the communicating is sending two messages of different orders of abstraction, and using one to deny the other; and (c) the victim cannot metacommunicate, cannot comment on this contradiction. Such contradictions become "reality," and over time the child may learn to metacommunicate by means of the most fantastic metaphors. The metaphorical and the literal become permanently confused, and the metaphorical is safer since it avoids direct comment and so does not put the victim on the spot. If the patient finally decides he is Napoleon, he is perfectly safe, because he has effectively accomplished what was previously not possible: he has left the field. The double bind cannot work any longer, because it is no longer he who is present, but "Napoleon." This is not, however, a game; if survival depends on being Napoleon, the victim will not be aware he is talking in metaphors, or that he is really not the historical Napoleon. Madness is not so simply the breakdown of the psyche. It is, in actual fact, an attempt to *salvage* the psyche.

Double-bind situations abound in psychopathology, and Bateson gives as a classic example the case of a visit made by a mother to her hospitalized son, who was recovering from a recent episode of acute schizophrenia.

> He was glad to see her [writes Bateson] and impulsively put his arm around her shoulders, whereupon she stiffened. He withdrew his arm and she asked, "Don't you love me any more?" He then blushed, and she said, "Dear, you must not be so easily embarrassed and afraid of your feelings." The patient was able to stay with her only a few minutes more and following her departure he assaulted a[n] orderly and was put in the tubs.

Clearly, continues Bateson, the result could have been avoided if the young man had been able to confront his mother with the fact that she became uncomfortable when he expressed affection for her. But years of intense dependency and training, going back to a time when he was a helpless infant, had set up a pattern that made this option impossible. Over the years he had learned, says Bateson, that "if I am to keep my tie to mother, I must not show her that I love her, but if I do not show her that I love her, then I will lose her." We see in this example a confusion of logical types. The child had learned that if he were to maintain his relationship with his mother,

> he must not discriminate accurately between orders of message.... As a result [he] must systematically distort his perception of metacommunicative signals.... He must deceive himself about his own internal state in order to support mother in her deception.

There is, then, no such thing as a schizophrenic person. There is only a schizophrenic *system*. The mother in such a system is in the position of controlling the child's definitions of its own messages, and (deutero-) teaches it a reality based on false discrimination of those messages. She also forbids the child to use the metacommunicative level, which is that level ordinarily used to correct our perception of messages, and without which such normal relationships become impossible. Yet modern psychiatry puts *the child* in the lockup, and lets the mother run free. A strong father might be able to intervene on the child's behalf early on, and in an extended family even an uncle or grandparent might save the situation. But madness has increased proportionally with the rise of the nuclear family, and it is typically the case in schizophrenogenic families that if the father (or the mother, if it is the man who is doing the double binding) were to step in to support the child, he would have to recognize the real nature of his own marriage—a recognition that would undo it. Schizophrenia is not a "disease" but a systemic network, a wonderland in which Alice is not free to tell the queen she is more than a little bit looney.

How *does* one escape from the double bind, then? On the individual level, at least, Bateson notes that the exit door is frequently creativity. In a later (1969) reflection on the double bind, and in his elaboration of "Learning III" in an article on "The Logical Categories of Learning and Communication" (1971), Bateson realized that schizophrenia was itself part of a larger system that he called the "trans-contextual syndrome." Jokes, which often involve the scrambling of the literal and the metaphorical, are a good example of this syndrome. They depend on a sudden condensation of logical types, a violation of the Russell-Whitehead theory ("A beggar told me he hadn't had a bite in three days, so I bit him"). There is, in fact, a double bind present in the etiology of a whole range of behavior—schizophrenia, humor, art, and poetry, for example—but the theory of the double bind does not formally distinguish between these activities or states of mind. There is no way to say whether a particular family will produce a clown or a schizophrenic, for example. Those whose life is enriched by trans-contextual gifts, says Bateson, or impoverished by them, have this in common: things are never just what they are. There is often, or even always, a "double take" involved, a symbolic level that distinguishes the Don Quixote from the Sancho Panza. Thus while the patient in the hospital canteen thinks "What can I do for you?" might be a sexual invitation, the comedian constructs a short story or TV situation comedy based on the very same confusion.

According to Bateson, the double bind is rooted in the theory of deutero-learning; trans-contextuality is a deutero-learned "trait." In work he did on mammalian communication in the 1960s, Bateson discovered that one could double bind a porpoise until schizophrenic symptoms were induced.[26] For example, first teach the animal a series of tricks (flips, somersaults, etc.) and deutero-teach the context—instrumental reward—by tossing it a fish every time it performs a trick. Then raise the ante: reward comes after three tricks are executed. Finally raise the ante to a level that assaults the entire Learning II pattern: reward the porpoise only after it invents a completely new trick. The creature goes through its entire repertoire, either one trick at a time or in sets of threes, and gets no fish. It keeps doing it, getting angrier, more vehement. Finally, it begins to go crazy, exhibit signs of extreme frustration or pain. What happened next in this particular experiment was completely unexpected: the porpoise's mind jumped to a higher logical type. It somehow realized that the new rule was, "Forget what you learned in Learning II; there is nothing sacred about it." The animal not only invented a new trick (for which it was immediately rewarded); it proceeded to perform four absolutely new capers that had never before been observed in this particular species of animal. The porpoise had become trans-contextual. It had broken through the double bind to what Bateson calls "Learning III." In Learning III, we literally rise to a new level of existence, and then look down and recall, perhaps fondly, our past consciousness, fraught with what we thought was irresolvable contradiction. "Oh yes," we may say; "*that's* what that was all about." But the formal etiology of creativity and schizophrenia remains the same. The principle is synergistic, says Bateson; "no amount of rigorous discourse of a given logical type can 'explain' phenomena of a higher type."[27]

A similar event occurs in the relationship between Zen master and student, in which the master poses an impossible problem, a double bind known as a "koan." Some of these are famous: "What is the sound of one hand clapping?," or "Show me your face before your parents conceived you." Bateson cites the one in which the master holds a stick over the pupil and cries, "If you say this stick is real, I'll hit you. If you say it isn't real, I'll hit you. If you keep quiet, I'll hit you"—a classic double bind. What constitutes the creative exit here is the nature of the metacommunication. The student can, for example, take the stick and break it in two, and the master might accept this response if he sees that the act reflects the student's own conceptual/emotive breakthrough.

In Learning III, the individual learns to change habits acquired in Learning II, the schismogenic habits that double bind us all. He learns that he is a creature who unconsciously achieves Learning II, or he learns to limit or direct his Learning II. Learning III is learning *about* Learning II, about your own "character" and world view. It is a freedom from the bondage of your own personality—an "awakening to ecstasy," as William Bateson once defined true education. This awakening necessarily involves a redefinition of the self, which is the product of one's previous deutero-learning. In fact, the self starts to take on a certain irrelevance; in Bateson's words, it ceases to "function as a nodal argument in the punctuation of experience." As we have seen, the journey can be dangerous. The problem of the self is so difficult that many psychotics will not use the first person singular in their speech. For others more fortunate, Bateson claims, there is a merger of personal identity with "all the processes of relationship in some vast ecology or aesthetics...." Or as Laing puts it in one of his most beautiful passages,

> True sanity entails in one way or another the dissolution of the normal ego, that false self competently adjusted to our alienated social reality; the emergence of the "inner" archetypal mediators of divine power, and through the death a rebirth, and the eventual reestablishment of a new kind of ego-functioning, the ego now being the servant of the divine, no longer its betrayer.[28]

It is here that we arrive at a crucial point, one that Laing has made over and over again in his work. The type of reasoning involved in schizophrenia is the same as that at work in art, poetry, humor, and even religious inspiration. The main difference is that the latter forms of trans-contextuality are more or less freely chosen, whereas the schizophrenic is caught up in a system not of his own making. But in formal terms, at least, schizophrenia represents a more highly developed form of consciousness than the varieties of Learning II which most of us have been taught. Yet what is the nature of this Learning II, at least on the official level? By and large, it is a charade. The modern reality-system requires allegiance to a logic that in actual practice has to be violated all the time. Western society has deutero-learned a Cartesian double bind and called it "reality"; it was precisely metacommunication (nuance, tacit knowing) that the Cartesian world view officially managed to destroy.[29] At the level of the dominant culture, we are supposed to believe that scientific knowledge is the only knowledge real or worth having; that analogue knowledge is nonexistent or inferior; and that fact and value have nothing to do with each other. None of this is true, but we are all required to live by these rules, and for the most part not to comment on them (except in books, I suppose). Yet where does insanity lie, in such a situation? As we saw in our discussion of Newton, we now live in a world turned upside down, a systemic double bind that has resulted in a kind of collective madness. The only way out of this double bind, it would seem, lies in rising to a new level of holistic consciousness which will facilitate new and healthy modes of behavior. Whereas a Cartesian analysis of modern knowledge and social problems winds up, as Nietzsche said, biting its own tail, a holistic analysis suggests that not all circles are vicious, and that there might be ways of stepping out of the present one. Bateson offers us a place to step, a non-Cartesian mode of *scientific* reasoning. For in the course of elaborating the nature of our schismogenic tensions, and the role of analogue knowledge in the transmission of information—discussions that necessarily include a critique of Cartesian dualism—he also developed a methodology that merges fact with value and erodes the barrier between science and art. This methodology is holistic rather than Cartesian, and as much intuitive as it is analytic. It is, to quote don Juan's admonition to Carlos Castaneda, "a path with a heart," and yet without any corresponding loss of rational clarity.

I have presented this chapter as an intellectual odyssey, Gregory Bateson's journey through a series of problems that are among the most fascinating any scientist or thinker might consider. His studies do not necessarily add up to a formal epistemology, but then the Scientific Revolution itself did not begin as a set of abstract principles, but rather as a series of investigations of diverse problems—falling bodies, planetary motion, light and color. Only much later did these investigations reveal a common methodology; the *ideology* of mechanism was more the work of Voltaire and La-place than of Descartes and Galileo. Yet in Bateson's case, it may not be premature to argue that the insights gleaned from studying Iatmul transvestism, learning theory, metacommunication, and schizophrenia do ultimately constitute an epistemological framework. Indeed, Bateson himself has elaborated this epistemology in some of his writings on cybernetic explanation. By its very nature, however, Bateson's epistemology resists linear explication. It is really a stance toward life and knowledge, a commitment rather than a formula. Like alchemy, his epistemology constitutes a praxis. In approaching a problem, Bateson sought to immerse himself in the world view being studied. His scientific sophistication notwithstanding, Bateson instinctively knew that most knowledge was analogue, that realities lay in wholes rather than parts, and that immersion (*mimesis*) rather than analytical dissection was the beginning of wisdom. To give a digital summary of his approach risks reifying it, and thereby rendering it worthless or even dangerous. "Let loose ends lead to their own ends," a friend of mine once wrote in one of her poems; and perhaps it would be best not to tie them up here. Certainly, no set of abstractions Bateson or I lay out in linear, discursive terms can grasp the larger noncognitive reality of life. But we live in this century, not the fourteenth or twenty-second, and for better or worse we are saddled with verbal-rational knowledge as the primary mode of exposition. It is with some ambivalence, then, that I turn to a linear and analytical exposition of Batesonian epistemology.

8

Tomorrow's Metaphysics (2)

Mere purposive rationality unaided by such phenomena as art, religion, dream and the like, is necessarily pathogenic and destructive of life; and . . . its virulence springs specifically from the circumstance that life depends upon interlocking *circuits* of contingency, while consciousness can see only such short arcs of such circuits as human purpose may direct. . . .

That is the sort of world we live in—a world of circuit structures—and love can survive only if wisdom (i.e., a sense of recognition of the fact of circuitry) has an effective voice.

—Gregory Bateson, "Style, Grace and
Information in Primitive Art" (1967)

A well-ordered humanism does not begin with itself, but puts things back in their place. It puts the world before life, life before man, and the respect of others before love of self.

This is the lesson that the people we call "savages" teach us: a lesson of modesty, decency and discretion in the face of a world that preceded our species and that will survive it.

—Claude Lévi-Strauss (1972 interview)

Batesonian epistemology is essentially an elaboration of an answer to a single question: What is Mind? As Bateson tells us in his Introduction to *Steps to an Ecology of Mind*, Western science has attempted "to build the bridge to the *wrong half* of the ancient dichotomy between form and substance."[1] Rather than explain mind (or Mind), Western science explained it away. But it is unlikely that we could start with substance (matter and motion) as the one explanatory principle, and deduce form, or mind, from it. In Bateson's way of thinking, Mind is—without being a religious principle or entelechy—every bit as real as matter.[2]

The reality of Mind in Bateson's world view gives his epistemology certain characteristics that are formally identical to alchemy and Aristotelianism. Fact and value are not split, nor are "inner" and "outer" separate realities. Quality is the issue, not quantity, and most phenomena are, at least in a special sense, alive. Yet there is one great difference between Bateson's work and all of those traditional epistemologies that are premised on the notion of a sacred unity: there is no "God" in his system. There is no animism, no *mana*, nothing of what we have called "original participation," because Mind is regarded as being immanent in the arrangement and behavior of phenomena, not inherent in matter itself. Thus, although there *is* such a thing as participation—we are not separate from the things around us—it does not exist in the "primitive" or premodern sense.

Earlier in this work, we delineated the differences between seventeenth-century science and its holistic predecessors. Before we proceed to an analysis of Batesonian epistemology, it will be useful to examine an outline of its differences from the Cartesian paradigm, as shown in Chart 2.[3]

We have commented on some of the above differences in Chapters 5 and 7, but most are not immediately obvious and will have to be spelled out in the discussion that follows. For now, I wish to

Chart 2. Comparison of Cartesian and Batesonian world views

World view of modern science	*World view of Batesonian holism*
No relationship between fact and value.	Fact and value inseparable.
Nature is known from the outside, and phenomena are examined in abstraction from their context (the experiment).	Nature is revealed in our relations with it, and phenomena can be known only in context (participant observation).
Goal is conscious, empirical control over nature.	Unconscious mind is primary; goal is wisdom, beauty, grace.
Descriptions are abstract, mathematical; only that which can be measured is real.	Descriptions are a mixture of the abstract and the concrete; quality takes precedence over quantity.
Mind is separate from body, subject is separate from object.	Mind/body, subject/object, are each two aspects of the same process.
Linear time, infinite progress; we can in principle know all of reality.	Circuitry (single variables in the system cannot be maximized); we cannot in principle know more than a fraction of reality.
Logic is either/or; emotions are epiphenomenal.	Logic is both/and (dialectical); the heart has precise algorithms.
Atomism:	Holism:
1. Only matter and motion are real.	1. Process, form, relationship are primary.
2. The whole is nothing more than the sum of its parts.	2. Wholes have properties that parts do not have.
3. Living systems are in principle reducible to inorganic matter; nature is ultimately dead.	3. Living systems, or Minds, are not reducible to their components; nature is alive.

point out that the differences involved are as profound as those that exist between science and alchemy, Sancho Panza and Don Quixote, or conventional sanity and Learning III. As Bateson himself once admitted, he had come a long way on the road from dualism, yet still thought in terms of an independent "I" and conceived of himself as a subject confronting objects. The statement is hardly surprising, for Bateson, or any other thinker writing about holism in the late twentieth century, remains a transitional figure. The fact that he retained the thought processes of our world is what enabled him to converse with us. But if Batesonian holism is indeed the mental framework of an emerging civilization, that civilization, once mature, will probably find our ways of thinking almost incomprehensible. It may even build museums of the history of science, in which visitors will have to turn their minds literally inside out in order to grasp what Galileo and Newton were trying to say.

Although Bateson learned about cybernetic theory in the course of the Macy Conferences, his understanding and elaboration of the theory developed in the context of concrete human situations. Curiously enough, Bateson chose to explicate the theory in an essay on alcoholism, "The Cybernetics of 'Self'" (1971), for his research revealed that the "theology" of Alcoholics Anonymous was virtually identical to cybernetic epistemology. Before summarizing Batesonian holism in formal terms, then, let us follow him through one more concrete investigation.[4]

It may at first seem strange that alcoholism could have anything whatever to do with epistemology. Yet as I hope is by now clear, philosophy and epistemology are not topics confined to academic circles. Wittingly or not, we all have a world view, and the alcoholic is no exception. As Bateson showed, our world view is, in effect, our "self," our "character," because it is the result of our deutero-learning. In the case of alcoholism, he discovered that in the oscillation between sobriety and intoxication, the alcoholic is actually switching back and forth between a Cartesian outlook and one that might be termed "pseudo-holistic." Bateson's point of departure was the attempt to uncover the dynamics of this oscillation.

With the exception of the efforts of Alcoholics Anonymous, all attempts to cure a drinking problem are based on the model of conscious self-control. The alcoholic is told to be strong, to resist temptation, to be "the master of my fate . . . the captain of my soul" (as William Ernest Henley wrote in "Invictus"). When sober, he agrees with these exhortations from his wife, his friends, his employer and others who supposedly seek to help him. The problem is that such advice represents pure Cartesianism; it is based on the assumption of a mind/body split. The mind (conscious awareness) is the "self" that is going to exert control over a weak and wayward body. But "cure" by self-control throws the entire situation into one of symmetrical schismogenesis: the conscious will is pitted in an all-out war against the rest of the personality. As in Freudian psychology, the unconscious (or body) is excluded from the self, and then seen as a collection of (evil) "forces" that the conscious self must struggle to resist. The alcoholic's resolution, "I will fight the bottle," "I will defeat demon rum," is a type of pride which derives directly from Cartesian dualism.

Why doesn't this approach work? As Bateson notes, the context of sobriety changes with achievement. There is a challenge involved in symmetrical struggle, and after the alcoholic manages to steer clear of liquor for a while, his motivation drops. Cartesian mind/body dualism, being schismogenic in nature, requires continual opposition in order to function, and that is the world view to which the alcoholic is committed. Not-drinking is no longer a challenge. But how about some "controlled drinking" (as AA mockingly calls it)? How about "just one drink"? This is indeed a challenge! And of course, he "falls off the wagon" and in short order is drunk once again.

What does the alcoholic perceive when drunk? At least in the initial stages of intoxication, a different personality emerges, and hence a different epistemology is ostensibly at work. In fact, the alcoholic switches, temporarily, from Cartesian dualism to what appears to be a holistic outlook. The mind abandons the attempt to control the body, and the struggle between them collapses, the result being, Bateson argues, a more correct state of mind. Getting drunk is a way of escaping from a set of cultural premises about the mind/body relationship which are in fact insane, but which society, in the form of husband, wife, friends, and employers, constantly reinforces. In a state of intoxication, however, the whole symmetrical contest drops to the ground, and the feelings that emerge are complementary. As the alcoholic begins to get drunk, he may feel close to his drinking buddies, to the world around him, and to his own self, which is no longer treating him in a punitive fashion. The abandonment of the struggle with himself and with the world

240

around him comes as a welcome relief. Cartesian dualism exhorted him to be "above it all," to be above being weak and human. Now, he seems more a part of the human scene. The psychology of contest (*agon,* in Greek, from which we get our word "agony") gives way to what appears to be the psychology of love.

The problem, however, is that this state of "love" is an illusion, almost as illusory as Cartesian dualism. In reality, the new state of mind is the pathology of submission. The alcoholic has but two strings on his guitar: rigidity (the "Invictus" posture) and collapse, or total vulnerability. He has no other behavior in his repertoire besides "triumphant" egotism and total capitulation. It was the genius of the founders of AA to recognize that these choices were two sides of the same coin, and that a third way might be possible.[5] This third way did capture the "truth" of the drunken state, the notion of surrender which is involved in it; but it was a surrender that conferred on the individual not maudlin impotence but power. In other words, it rendered him *active* in the world; it was not an illusory state, or a short circuit, but a circuit that was dynamic and continuous.

How did AA manage to do this? Consider the first two steps of its program: (1) we admitted we were powerless over alcohol—that our lives had become unmanageable—and (2) we came to believe that a Power greater than ourselves could restore us to sanity. The first step undercuts Cartesian dualism in a single stroke. That dualism pits "sober" mind versus "alcoholic" body, implying that demon rum is somehow *outside* the personality, outside the body. The "decent," "pure," "noble" conscious will—which is "in here"—is trying to control the "weak," "dirty" alcoholic body— which is "out there." Once the alcoholic comes to an AA meeting and says to the group, "My name is John Doe, and I am an alcoholic," he places the alcoholism within his self. The total personality has admitted to being alcoholic. It is no longer a case of the alcoholism being "out there." Once you surrender, admit that you are powerless over the bottle, abandon the sloganeering of "Invictus"—which AA in fact uses as a point of ridicule—the symmetrical battle evaporates, without your getting drunk.

AA's second principle provides the basis for an alternative epistemology that is genuinely holistic. By definition, you can only be in a dependent relationship to a Higher Power. This admission seems like a surrender, says Bateson, but in fact it is really a change in epistemology, and therefore in character or personality. This

241

Higher Power—"God as you understand Him to be," as AA says—is of course the unconscious mind, but is more than this as well. It is also your social reality, the other members of AA, and the struggle that their lives represent. The individual ego (conscious will) leaves the field in favor of a more mature form of self; one that is both intra- and inter-personal. Such a surrender is not a collapse, but a renewal. For the alcoholic who has finally "hit bottom," as AA calls it, the first two steps of the AA program in effect constitute Learning III, and the alcoholic frequently experiences them as a religious conversion.

What does this analysis have to do with cybernetic theory? The metaphysics of Western science deals with atoms, with single individuals, and with causes that are direct, conscious, and empirical. The Cartesian paradigm would, for example, isolate the alcoholic and attempt to ascertain the "cause" producing the undesirable "effect." The theory is one of direct linear influence, based on the model of seventeenth-century impact physics in which mind is viewed as explicitly conscious and external to matter. In Descartes' view of things, God is outside it all; He merely set the whole arrangement in motion. Similarly, the balls on a billiard table have no inherent mind; mind comes *to* them in the form of a person with a cue stick.

In cybernetic theory, on the other hand, the unit to be considered is the whole system, not this or that individual component. Consider the ensemble of a steam engine plus its control unit, commonly known as a "governor." As in the case of a thermostat controlling the temperature of a house, the governor is set in terms of an ideal—in this case the optimal running speed of the engine. Should the actual speed fall much below the ideal, the armature slows down until the fuel supply is triggered, bringing the speed up to "normal." Conversely, if the engine starts moving too fast, the swinging armature triggers the brake, and the system is once again brought into line. But what influences the governor, or self-corrective feedback mechanism, is not some Cartesian impact, some billiard ball or concrete entity, but only information. And a "bit" of information, also known as an "idea," Bateson defines as "a difference which makes a difference." In other words, the engine, governor, fuel supply, brake, locomotive, and other components form a complex causal circuit. A change, or difference, in the operation of any single component is felt throughout the system, and the system reacts with something that might be termed awareness, if not consciousness. In this sense, it is alive. It possesses mental characteristics, and can be regarded as a mind (Mind) of some sort. We assert, writes Bateson, "that *any* ongoing ensemble of events and objects which has the appropriate complexity of causal circuits and the appropriate energy relations will surely show mental characteristics." In other words, it will make comparisons (be responsive to differences), process information, be self-corrective towards certain optima, and so on. Furthermore, adds Bateson, "no part of such an internally interactive system can have unilateral control over the remainder or over any other part. The mental characteristics are inherent or immanent in the ensemble as a *whole.*"

Now a mental system, a Mind, can exhibit one of three possible types of behavior: self-correction (also called steady state), oscillation, or runaway. Here is the link between schismogenesis and cybernetic theory. *A schismogenic situation is one without a governor; the system is constantly slipping into runaway.* In a self-corrective system, the results of past actions are fed back into the system, and this new bit of information then travels around the circuit, enabling the system to maintain something near to its ideal, or optimal state. A runaway system, on the other hand, becomes increasingly distorted over time, because the feedback is positive, rather than negative or self-corrective. Addiction is the perfect example of a runaway system. The heroin addict needs an increasingly larger fix; the sugar addict finds that the more pastry he eats, the more pastry he wants; the imperialist power starts out seeking particular foreign markets, and eventually winds up trying to police the globe.

Although the ethical implications of these alternatives will be discussed later, it might be appropriate to point out an obvious corollary of this cybernetic analysis. Given the fact that schismogenesis is so pervasive a phenomenon in Western culture, we are forced to conclude that the institutions and individuals of that culture are in various degrees of runaway. Addiction, in one form or another, characterizes every aspect of industrial society, down to the lives of individual members. Dependence on alcohol (food, drugs, tobacco . . .) is not formally different from dependence on prestige, career achievement, world influence, wealth, the need to build more ingenious bombs, or the need to exercise conscious control over everything. Any system that maximizes certain variables, violating the natural steady-state conditions that would *op-*

timize these variables, is by definition in runaway, and ultimately, it has no more chance of survival than an alcoholic or a steam engine without a governor. Unless such a system abandons its epistemology, it will hit bottom or burn out—a realization that is now dawning on many individuals in Western society. There is no escaping self-corrective feedback, even if it takes the form of the total disintegration of the entire culture. A mental system cannot remain in permanent runaway, cannot maximize variables and also retain the characteristics of Mind. It *loses* its Mind; it dies. On the individual level, we experience cirrhosis, heart attack, cancer, schizophrenia, and what has to be called living death. The ethics of the system are implicit in its epistemology.

The example of alcoholism enables us to understand the status of the "self," or conventional "mind" (Cartesian ego), in cybernetic theory. As we have noted, Bateson claims that the mental characteristics of a cybernetic system are immanent not in some particular part, but in the system as a whole. *The conscious mind, or "self," is an arc in a larger circuit,* and the behavior of any organism will not have the same limits as the self. Alcoholic "pride," or determined sobriety, is the attempt to maximize the variable called conscious mind, to have this little arc somehow get control over the entire circuit. Such pride is the foolishness of "Invictus," at least as applied to addiction, for there is more to a steam engine than its governor. Being drunk, or in a state of collapse, is a shortcut to complementarity, and a short-term solution. The wisdom of AA is to switch the system from runaway to self-correction by introducing complementary elements into a symmetrical situation, and introducing them in such a way that the resulting recognition of circuitry becomes self-sustaining.

Bateson uses the example of a man chopping down a tree to demonstrate the circuitous nature of Mind. According to the Cartesian paradigm, only the man's brain possesses consciousness: the tree is of course alive, but it is not (in this view) a mental system of any sort, and the axe itself is dead. The interaction is causal and linear: man takes axe and operates on tree trunk. He may say to himself, as he does this, "I am cutting down this tree," the thesis being that there is a single entity, "I," the self, which is undertaking purposive action upon a single object. The fallacy here is that mind is introduced in the word "I," but is restricted to the man, whereas the tree is reified, seen as an object. But the mind winds up being reified also; for since the self acted upon the axe, which then acted upon the tree—a perfect application of Cartesian impact physics—the self must also be a thing, and therefore dead. Moreover, when we try to localize the self in such a system, we find we cannot do so. In another Batesonian example, that of a blind man making his way down the street with the help of a stick, there is no way to say where his self begins or ends. Isn't the stick really part of his self? He is not simply acting upon it, as an object, which then acts upon the pavement. The stick is really a *pathway* to the pavement, to his environment. But where does the pathway end? At the handle? The tip? Halfway up the stick? "These questions," writes Bateson, "are nonsense, because the stick is a pathway along which differences are transmitted under transformation, so that to draw a delimiting line *across* this pathway is to cut off a part of the systemic circuit which determines the blind man's locomotion." The mental system of the blind man—or any of us—does not end at the fingertips. To explain the man's locomotion, says Bateson, you need the street, the stick, and the man; and the stick becomes irrelevant only when he sits down and puts it aside.

The same argument applies to the man and the axe. Each stroke is modified according to the shape of the cut left by the previous stroke. There is no "self" "in here" cutting down a tree "out there"; rather, a relationship is occurring, a systemic circuit, a Mind. The whole situation is alive, not just the man, and this life is immanent in the circuit, not transcendent. The mind may indeed be the man's frontal lobes, but the larger issue is the Mind, which in this case is "tree-eyes-brain-muscles-axe-stroke-tree." More precisely, what is going around the circuit is *information*: differences in tree/differences in retina/differences in movement of axe/differences in tree, and so on. This circuit of information is the Mind, the self-corrective unit, now seen to be a network of pathways which is not bounded by purposive consciousness, or by the skin, but extended to include the pathways of all unconscious thought, and all the external pathways along which information can travel.

Clearly then, large parts of the thinking network lie outside the body, and the statement that Mind is immanent in the body, which I made (more or less) in Chapter 6, can now be seen as a stepping-stone to this discussion. Tacit knowing is not merely a physiological phenomenon. The study of alcoholism, schizophrenia, and deutero-learning has demonstrated that such phenomena are not matters of individual psychology, but of Minds, or systems, not bounded by the skin of the participants. "Self" is a false reification of a small part of a larger informational network, and we make the same mistake when we introduce such reification into the relationship between a man and the tree he is cutting or into any other interaction or understanding we might have with, or of, "inert" objects. In terms of a cybernetic interpretation of what constitutes an event, and a Mind, the world view of Galileo and Newton is literally nonsensical, and the world view of the alchemists, which was posited on the absence of a subject/object distinction, profoundly correct.

We are now ready to consider cybernetic epistemology as a formal system, which can be done by making explicit those items that can be regarded as criteria of Mind, or mental system. These are as follows:[6]

(1) There is an aggregate of interacting parts, and the interaction is triggered by differences.

(2) These differences are not ones of substance, space, or time. They are nonlocatable.

(3) The differences and transforms (coded versions) of differences are transmitted along closed loops, or networks of pathways; the system is circular or more complex.

(4) Many events within the system have their own sources of energy, that is, they are energized by the respondent part, not by impact from the part that triggers the response.

Before discussing each of these points in turn, let us note that according to this set of criteria, a social or political structure, a river, and a forest are all alive, and possess Mind. Each has its own energy sources, forms an interlocking aggregate, acts self-correctively, and has the potential for runaway. Each knows how to grow, how to take care of itself, and should these processes fail, how to die. As Bateson says, all the phenomena we call thought, learning, evolution, ecology, and life occur only in systems that satisfy these criteria. Let us elaborate on them briefly.

(1) There is an aggregate of interacting parts, and the interaction is triggered by differences. We have already discussed this criterion in the case of the steam engine, the man chopping down the tree, and the blind man with the stick. In each case, information—differences that make a difference—circulates through the system. The blind man suddenly slows down as the stick tells him he is at the edge of a curb; a whole different process is set in motion as he feels his way across the street. Differences in muscles make dif-

ferences in movements make differences in retina make differences in brain make differences in exposed surface of the tree trunk, and such differences circulate around the system of man-cutting-down-tree, influencing one another in a continual, changing cycle.

Furthermore, parts of the aggregate—the tree for example—may also satisfy these conditions, in which case they are sub-Minds. But there is always a sublevel that is *not* alive, the axe by itself for instance. The explanation of mental phenomena is thus never supernatural. Mind always resides in the interaction of multiple parts that may, of themselves, not satisfy the criteria of Mind.

(2) These differences are not ones of substance, space, or time; they are nonlocatable. This statement represents another way of rejecting the Cartesian impact physics model, or linear causality. The model certainly works for interacting billiard balls, or Newtonian studies of force and acceleration, but once a living observer is admitted to be part of such cases, the cause of events is no longer a force or an impact. An observer, or receiver, responds to a difference or a change in a relationship, and this difference cannot be located in any conventional sense.

Consider, for example, the difference between the blackness of the ink in this sentence and the whiteness of the paper on which it is printed. Few people would deny that there is a real difference here. But where is it? The difference is not in the ink; it is not in the white background; it is not in the "edge," or outline, between them, which is after all a collection of mathematical curves, possessing no dimension. Nor is it in your mind, any more than the ink or the paper are actually in your mind. A difference is not a thing or event. It has no dimension, any more than do such abstractions as congruence or symmetry. Yet it exists, and to complicate matters further, nothing—that which is *not*—can be a cause. As Bateson points out, the letter you do not write can get an angry reply; the tax form you fail to submit can get you in trouble. There is no parallel here to the world of impact physics, where impacts are causes, where real things must have dimension, and where it takes a "thing" to have an effect.

(3) The differences and transforms (coded versions) of differences are transmitted along closed loops or networks of pathways; the system is circular or more complex. We have, essentially, discussed this criterion in our analysis of the feedback process. Another way of stating it might be to say that the system is self-corrective in the direction of homeostasis and/or runaway, and that

247

self-correctiveness implies trial and error behavior. Nonliving things maintain a passive existence; living entities, or Minds, escape change *through* change, or more precisely, by incorporating continual change into themselves. Nature, says Bateson, accepts ephemeral change in favor of long-term stability. The bamboo reed bends in the wind so as to return to its original position when the wind dies down, and the tightrope walker shifts his or her weight continually to avoid falling off the high wire. Even runaway systems contain seeds of self-correction. Symmetrical tensions run so high among the Iatmul that complementary naven behavior is almost constantly being triggered. The alcoholic usually comes to AA when he or she has finally hit bottom. Marx's argument that capitalism was, by its very nature, digging its own grave, is also an example of cybernetic thinking; and phenomena such as famine, epidemics, and wars might be regarded as extreme cases of nature's attempt to preserve homeostasis. The current collapse of industrial society may well be the planet's way of avoiding a larger death.

(4) Many events within the system have their own sources of energy, that is, they are energized by the respondent part, not by impact from the part that triggers the response. This criterion is another way of saying that living systems are self-actualizing, that they are subjects rather than objects. The reaction of a dog that you kick comes from the animal's own metabolism; the two feet it might have traveled from the force of your kick is less significant than the dog's subsequent response, which might include taking a chunk out of your leg.

Given these criteria of Mind, the next obvious question is: How do we know the world; which is to say, other Minds? On the Cartesian model, we know a phenomenon by breaking it into its simplest components and then recombining them. Enough has been said already to indicate how fallacious this atomistic approach really is. In fact, in terms of cybernetic theory, Cartesian analysis is a way of *not* knowing most phenomena, because Mind can only be characteristic of an (interacting) aggregate. Meaning is virtually synonymous with context. Abstract a thing from its context (a ray of light for example) and the situation becomes meaningless, although perhaps mathematically precise.

In cybernetic theory, then, we can know something only in context, in its relation with other things.[7] In addition to "context," Bateson uses other words to denote "meaning," and these are "redundancy," "pattern," and "coding." The circulation of information involves the reduction of randomness, a process that can also be called the creation of negative entropy (entropy is the measure of randomness of a system). If something is redundant, if it possesses a definite pattern, then it is not random and constitutes a source of information. Communication is thus the creation of redundancy, and redundancy is the central epistemological concept in cybernetic theory, which is the science of messages. It is interesting to note, once again, that this concept is an advanced form of an idea first advanced by William Bateson, namely the "undulatory hypothesis" (see Chapter 7). Redundancy is an undulatory hypothesis; both terms are derived from the Latin word *unda*, wave. A re-dundant situation is one in which wave after wave of similar or identical information washes over us. The holistic outlook of both Batesons is rooted in the notion that we know the world through redundancy.

Gregory Bateson takes the following definition as his paradigm for knowing:

> Any aggregate of events or objects (e.g., a sequence of phonemes, a painting, or a frog, or a culture) shall be said to contain "redundancy" or "pattern" if the aggregate can be divided in any way by a "slash mark," such that an observer perceiving only what is on one side of the slash mark can *guess,* with better than random success, what is on the other side of the slash mark. We may say that what is on one side of the slash contains *information* or has *meaning* about what is on the other side.

Much of the information we absorb is digital in nature, usually spoken or written. If I say "on the one hand," you know there is another hand lurking somewhere in the wings, and you know what this means. Clichés are redundant to the point of rigidity. The term itself originally applied to blocks of typeface which were glued together by printers because they occurred so often in published work. The English language is also redundant at the level of individual letters. Given a letter T in a piece of prose, we know that the next letter is almost certainly H, R, W, or a vowel (including Y). Words like "tsetse" and "tmesis" tend to catch our attention, for their spelling is less redundant than the spelling of "than" or "the."

Most of the information we take in, however, is analogue, or iconic. As I walk down a street alongside a large building, unable to see around the corner, I expect to find right angles in both street and building as I make the turn. This is in fact the equivalent of a cliché. If, however, I frequently fell down a mine shaft as I turned such a corner, the situation would be so lacking in meaning that I would never leave my house. Clichés, as we know, are safe.

The entire world of metacommunication also has this structure. From a gesture or tone of voice we guess across the slash mark what is really meant:

"I love you" (impatient tone of voice)/Rejection

For this same reason , as we have already seen, there is no such *thing* as an "ethos" or "character." "Dependency," "hostility," and so on are patterns, and from a person's behavior we guess their state of mind, that is, across the slash mark. A redundant behavior pattern, such as the ones Freud records in his list of human defense mechanisms, or those that Eric Berne reproduces in *Games People Play*, does tend to become like a cliché, and lead us to think of the pattern as a concrete item, a "trait."

By contrast, one reason we enjoy a demonstration of skill, whether the performer is playing the piano or juggling balls while balancing on a monocycle, is that we understand instinctively that skill is a coding of unconscious information; a coding that is, unlike a cliché, difficult to achieve. The gracefulnes of the act reveals a certain level of psychic integration, which, understandably, fascinates us. In such cases the redundancy takes this form:

Performance/conscious-unconscious relationship

It is this type of redundancy that enables us, for example, to appreciate the art of cultures completely different from our own. We can somehow feel the degree of authenticity, or the degree of conscious-unconscious integration, from the skill or performance shown.

It is at this point that the principle of incompleteness, or indeterminacy, as is present in quantum mechanics, becomes crucial. In Chapter 5, I pointed out Bateson's essential agreement with this notion, as opposed to the Freudian or Cartesian notion that everything can, in principle, be known. Our discussion of redundancy shows us that if all tacit knowing could be made explicit, all unconscious information be made conscious, there would not be anything that was not a cliché. Everything would be totally stylized, totally formalistic, and thus also totally random—meaningless. The general structure of communication, of meaning, is necessarily part-for-whole; and to have it all spelled out, to erase the slash mark altogether by making everything redundant, erases the possibility of creating redundancy at all. It is not without good reason that Polyani calls the attempt to do this, to make everything explicit, a program for reducing the human race to a state of "voluntary imbecility."[8]

The principle of incompleteness gives Batesonian holism its real power, turning what is a weakness in conventional science into a source of strength. It says, in a nutshell, that mind is not Mind, nor, in principle, can it ever be so. It argues that by definition, tacit knowing can never be rationally expressed. But we can recognize its existence, we can work with it in our attempt to know the world, and in fact we must do so because circuitry, in the cybernetic sense, is the way reality is structured.

At the time of his research for *Naven*, Bateson had seen incompleteness as a problem. In particular, he felt that "ethos" was too intangible (analogue) a thing to grasp. The real weakness in his study, he stated in the 1936 Epilogue, was not so much his own theoretical treatment as the absence of any science of tacit knowing. "Until we devise techniques for the proper recording and analysis of human posture, gesture, intonation, laughter, etc.," he wrote, "we shall have to be content with journalistic sketches of the 'tone' of behaviour."[9] This lacuna continued to confront him in each area that he studied. Deutero-learning was largely a matter of analogue cues. Schizophrenia pivoted on disturbances in metacommunication. On the surface, a science of analogue behaviour seemed to be precisely what was needed for the resolution of such problems. In his Balinese studies, Bateson tried to fill this gap by a very innovative use of field photography; and Jurgen Ruesch (a later coworker) and other researchers went on to make the whole subject of kinesics and paralinguistics into a separate academic discipline.[10] By and large, however, Bateson own work ultimately moved in a very different direction. He not only came to the conclusion that it would be unwise to try to illuminate fully this sort of unconscious information, but that, in principle, it could not be done; that analogue and digital modes of knowing were not really mutually translatable. He became convinced that this gap in

251

our knowledge was not something for science to "solve," but that it constituted a scientific fact of life. The situation is similar to the relationship between figure and ground in gestalt psychology. They are not symmetrical, their relationship is not one of simple opposition. Digital knowledge makes itself evident by "punctuating" analogue knowledge; the latter is hardly dependent upon the former for its existence. Analogue knowledge is pervasive, vast; it is the ground of perception and cognition. In premodern culture, the digital (when it did exist) was the instrument of the analogue. After the Scientific Revolution, the analogue became the instrument of the digital, or was suppressed by the latter entirely, to the extent that such suppression was possible. This distortion, which Freud exalted as the hallmark of health, Bateson saw as the crux of our contemporary difficulties. Converting all id to ego, or trying to spell out cardiac algorithms in cognitive-rational terms, was a continuation of the program of the Scientific Revolution and its distorted epistemology. In a healthy epistemology, the two modes of knowing would be used to nourish and complement one another. Our culture, with its heavy emphasis on the digital, could restore such a complementary relationship only by recovering what it once knew about archaic modes of thought. But to try to elaborate these modes in empirical-conscious terms was, Bateson concluded, in fact to destroy them in the name of understanding them.[11]

To understand this point more clearly, consider the popular theory that language replaced earlier iconic systems of communication in the history of human evolution. Once messages could be articulated verbally or in writing, communication by way of signs, drum beats, and so on simply fell into disuse. The problem with this theory, says Bateson, is that analogue communication, including human kinesics, has in fact become richer. Rather than being discarded, these archaic modes have themselves evolved. We now have Cubism as well as cave paintings, ballet as well as rain dances. This is not to argue that modern forms are more sophisticated than archaic ones, for evolution is not synonymous with progress. But our repertoire of communication has become more sophisticated with the passage of centuries; and the evolution of iconic communication suggests that such communication serves functions somewhat different from those served by language, and that it was never intended to be replaced by the latter. To translate kinesics into words (specifically, prose) says Bateson, falsifies things, because such translation must give the appearance of conscious intent to a message that is unconscious and involuntary.

Since the essence of an unconscious message is that it *is* unconscious, that there *is* such a thing as unconscious communication,
the translation necessarily destroys the nature of the message, and
thus the message itself. Freud's theory of repression, that the unconscious is the repository of painful memories, is a very confused
theory in that much of what exists in the unconscious was always
there. According to Freud's view, poetry would be a type of distorted prose, whereas the truth is that prose is poetry which has
been converted into a "logical" presentation.

I have already noted Bateson's example of the hypothetical television set that reports on its own internal workings as an illustration of the limits of consciousness. We see the paradox at once: it is
as though I were to say to you, "Speak to me about what you are
speaking as you are speaking it." In order for the television to
report on the workings that make possible that very report,
another unit would have to be added to it. But since this new unit
could not report on *its* own workings, a unit would have to be
added to that, and so on. One would soon confront an infinite
regress, a set of Chinese puzzle boxes. The attempt of the conscious mind to explicate its own mode of operation involves the
same sort of paradox. But there is an additional confusion that
derives from the different types of communication involved. As
already noted, all analogue communication is an exercise in communication about the species of the unconscious mind, about the
way it itself works. But the unconscious mind is no more able
logically to do this than the conscious mind; it can only show what
it is about by working in the way it does, that is, according to the
rules of primary process. A skilled performance is the deliberate
attempt to display the nature of spontaneous, nondeliberate behavior. Thus Bateson suggests that the usual interpretation of a
remark attributed to Isadora Duncan is wrong. She supposedly
said: "If I could tell you what it meant, there would be no point in
dancing it." As Bateson says, the common interpretation is something like, "There would then be no point in dancing it, because I
could tell it to you, quicker and with less ambiguity, in words."
This interpretation is all of a piece with the program of making the
unconscious totally explicit. There is, says Bateson, another possible interpretation, one which Isadora probably had in mind:

> If the message were the sort of message that could be communicated
> in words, there would be no point in dancing it, but it is not that sort
> of message. It is, in fact, precisely the sort of message which would be

falsified if communicated in words, because the use of words (other than poetry) would imply that this is a fully conscious and voluntary message, and this would be simply untrue.

Digital knowledge can only communicate conscious intent. If the message itself is, "There is a species of knowledge that is not conscious or purposive," its expression in digital terms is necessarily the falsification of the message rather than the expression of it. "Let me dance to you an aspect of tacit knowing," Isadora is saying; let me show you what life is really about. It is not merely that what we consciously know is only a fraction of reality, but that incompleteness of knowledge is the source of knowledge itself (if I could dance this book, I wouldn't have to write it). If Western science could somehow achieve its program of total certainty, at that very moment it would know nothing at all.[12]

As I stated at the end of Chapter 7, the Batesonian paradigm cannot genuinely be formulated in a digital fashion, any more than the alchemical paradigm can. Both recognize that incompleteness is inevitably part of the process of reality itself. The closest we can come to formulating Bateson's paradigm is through the study both of specific examples (as we have done) and the method of his investigation. We thus have holistic answers to questions such as: What is schizophrenia? What is alcoholism? How do mammals learn? It seems to me that the holistic approach can be extended to questions such as, What are light and color? What is electricity? Why do objects fall to earth? Our present mechanistic answers to such questions are clearly insufficient, especially because they incorrectly leave the observer and his entire range of analogue/affective behavior out of the investigation. The research undertaken by a future holistic science would take incompleteness and circuitry as axioms; would seek to uncover the cybernetic properties of a situation, while including the human investigator in the circuit being studied; would show how the analogue and digital patterns interlock; and would consider a specific piece of research "finished" when the nature of the Mind present in the situation had been satisfactorily explicated. Ultimately, the explication may not take a digital form at all, but instead appear as a videotape, a mime, or a book filled with collage. The goal of the research would be to deepen our relationship to nature by demonstrating its beauty—as was, for example, Kepler's purpose in his study of planetary harmony. The end result would be a better orientation of

254

ourselves in the cosmos. The notion of *mastering* the cosmos would, in a society built on holistic thought, make schoolchildren giggle, and produce blank, uncomprehending stares in adults.

What might a holistic society be like? I have argued that the horror of the modern landscape can at least partly be traced to the Cartesian paradigm, and have suggested that its insistence on a split between fact and value, or epistemology and ethics, is particularly to blame. For modern science, "What can I know?" and "How shall I live?" are totally unrelated questions. Science cannot, supposedly, tell us what the good life is. Of course, this modesty is highly suspect: "value-free" is itself a value judgment, amorality a certain species of morality. In Batesonian holism, as in the Hermetic world view and other systems of premodern thought, this false modesty is happily absent. A certain ethic is directly implicated in Bateson's epistemology; or, as he himself puts it, "the ethics of optima and the ethics of maxima are totally different ethical systems."[13] Since we already know a great deal about the ethics of maxima, of trying to master the environment, it will be necessary to conclude this chapter with an examination of the ethics of optima, and the sort of society that might be congruent with the holistic or cybernetic vision (I shall have more to say about this matter in specifically political terms in Chapter 9).

Much of the ethics implicit in Bateson's world view emerges quite explicitly when his epistemology is applied to living systems. Although it would be too much of a digression to discuss Bateson's writings on biology, including his radical revision of Darwin's evolutionary theory, we can nevertheless point to four crucial themes in that body of work which have immediate ethical implications:

(1) All living systems are homeostatic, that is, they seek to optimize rather than maximize certain variables.

(2) What we have identified as the unit of Mind turns out to be identical to the unit of evolutionary survival.

(3) There is a fundamental physiological distinction between addiction and acclimation.

(4) Species diversity is preferable to species homogeneity. Let us consider each of these themes in turn.

Although it is not at first evident, points (1) and (2) turn out to be variations on the cybernetic themes of circuitry and incompleteness. To review these notions briefly, we might think of Mind as a circle intersected by a plane, such that most of the circle is below the plane and only a small arc remains visible. The Cartesian paradigm holds that this visible portion—mind, or conscious awareness—is the sum total of nonmaterial reality. (Alternatively, it is seen as epiphenomenal, reducible to matter, and thus not really even there.) In the Freudian version of this paradigm, the larger reality is recognized, but regarded as dangerous, and the goal of the human system is to maximize the control exerted by the arc to include the entire circle. Ultimately, the Freudian goal is to transform the entire portion below the plane into the type of thinking which exists above the plane; in short, to eradicate it.

In Jungian, Reichian, or Batesonian terms, the goal of the human system is to make this plane highly osmotic. For Jung, what is below the plane is the unconscious. For Reich, it is the body, the true body, ecstatic and unarmored. For Bateson, it is tacit knowing, the complex set of informational pathways (including the social and natural environment) which constitute any system characterized by Mind. For all three, to make the plane completely permeable is to achieve wholeness, or "grace." This achievement does not dissolve the ego, the visible arc, but rather puts it in context, sees it as a small portion of a larger Self. Wisdom, in Bateson's terms, is the recognition of circuitry, the recognition of the limits of conscious control. The part can never know the whole, but only—if wisdom prevails—put itself at its service.

The relation between these notions and point (1) is that the circuit is a homeostatic system, and should there be an attempt to maximize any single variable, including the one alternatively called "mind," "conscious awareness," or "purposive rationality," the system will go into runaway, destroying itself and its immediate environment in the process.[14] Physiological systems are inherently structured in this way. The human body, for example, needs only so much calcium. We do not say, "the more calcium I have in my body, the better," because we understand that past a certain point any chemical element becomes toxic to an organism, no matter how essential it is to its health. In biological terms, the value systems of living entities are always biased toward optimization.

Somehow, although Western society is aware of this truth in biological terms, it pays very little attention to it otherwise. We cannot have too much rational consciousness, too much profit or power, too many accomplishments, too gross a Gross National Product. In cybernetic terms, such thinking is self-destructive, unwise. Bateson notes that the cybernetic nature of the self gets obscured to the extent that we become mesmerized by considerations of purpose. Cybernetics has a significant insight into the nature of stability and change. It understands that change is part of the effort to maintain stability. Purposive behavior, or maximizing behavior, on the other hand, limits the awareness of circuitry and complexity and leads to progressive change—runaway.

What is an example of an optimizing system, one that understands the facts of circuitry, and successfully preserves its own homeostasis? In response to this question, Bateson draws on his knowledge of Bali. The Balinese recognize that stability requires change and flexibility, and have created a society that Bateson appropriately calls "steady state." The emphasis is on balance—no variable is deliberately maximized—and the ethics of the situation is "karmic," that is, it obeys a law of nonlinear cause and effect, especially with respect to the environment. As Bateson puts it, "lack of systemic wisdom is always punished." If you fight the ecology of a system, you lose—especially when you "win."

Our second point, that the unit of Mind is identical to the unit of evolutionary survival, is a variation on point (1). In cybernetic theory the circuit is not a single individual, but the network of relations in which he or she is embedded. Of course, any living organism satisfies Bateson's criteria of Mind, but there are always Minds within Minds (see Plate 19). A man by himself is a Mind, but once he picks up an axe and starts to chop down a tree, he is part of a larger Mind. The forest around him is a larger Mind still, and so on. In this series of hierarchical levels, the homeostasis of the largest unit must be the issue, as the evolution of species has demonstrated. The species that cannot adapt to changes in its environment becomes extinct. Thus "person" or "organism" has to be seen as a sub-Mind, not as an independent unit. Western individualism is based on a confusion between sub-Mind and Mind. It regards the human mind as the only mind around, free to maximize any variables it chooses, free to ignore the homeostasis of the larger unit. Batesonian ethics, in contrast, is based on *relationship*, the recognition of the complex network of pathways. The posture of "Invictus," of the independent self so dear to Western thought, is foreign to Bateson's way of thinking. He regards this independence as a superficial freedom that, once surrendered, reveals a different sort of freedom which is much more comprehensive. Thus he holds that Darwin's theory of natural selection was correct—the fittest do survive—but that Darwin misidentified the unit of survival. "The unit of survival," writes Bateson, "—either

Plate 19. M. C. Escher, *Three Worlds* (1955). Escher Foundation, Haags Gemeentemuseum, The Hague.

in ethics or evolution—is not the organism or the species but the largest system or 'power' within which the creature lives. If the creature destroys its environment, it destroys itself."

Mind, he continues, is immanent in the ecosystem, in the total evolutionary structure. "Survival" means something different if it is extended to include the system of ideas in a larger circuit, not just the continuation of something bounded by skin. The ecosystem, in short, is *rational* (in the sense of being reasonable), and there is no violating its rules without suffering certain consequences. In pitting his own survival against the survival of the rest of the ecosystem, in adopting the Baconian program of technological mastery, Western man has managed, in a mere three centuries, to throw his own survival into question. The true unit of survival, and of Mind, is not organism or species, but organism + environment, species + environment. If you choose the wrong unit, and believe it is somehow all right to pollute Lake Erie until it loses its Mind, then you will go a little insane yourself, because you are a sub-Mind in a larger Mind that you have driven a bit crazy. In other words, says Bateson, the resulting insanity becomes part of *your* thought and experience, and there are clear limits to how many times you can create such situations before the planet decides to render you extinct in order to save itself. The Judeo-Christian tradition sees us as masters of the household. Batesonian holism sees us as guests in nature's home.

To conclude points (1) and (2), then, the world view advocated by Bateson, in both its ethics *and* its epistemology, is in direct contrast with secular humanism, the Renaissance tradition of individual achievement and mastery over nature. Bateson regards this sort of arrogance as completely unscientific. His own humanism, like that of Claude Lévi-Strauss, is based on the lessons of myths, the wisdom of "primitives," and the archaic algorithms of the heart. It is not opposed to the scientific intellect, but only to the inability of that world view to locate itself in a larger context.

The third point, that of the basic physiological distinction between acclimation and addiction, describes what happens when a homeostatic system is disturbed.[15] Bateson illustrates acclimation as follows:

> If a man is moved from sea level to 10,000 feet, he may begin to pant and his heart may race. But these first changes are swiftly reversible: if he descends the same day, they will disappear immediately. If, however, he remains at the high altitude, a second line of defence appears.

He will become slowly acclimated as a result of complex physiological changes. His heart will cease to race, and he will no longer pant unless he undertakes some special exertion. If now he returns to sea level, the characteristics of the second line of defence will disappear rather slowly and he may even experience some discomfort.

As Bateson points out, the process of acclimation manifests an impressive similarity to learning, especially Learning II. In fact, acclimation is a special case of the latter. The system becomes dependent upon the continual presence of a factor that was initially regarded as extraneous; it deutero-learns a new context. The same thing is true of addiction, but the factor in that case is actually inimical to the survival of the system, and—as we have seen in the case of alcoholism—reversibility is impossible without undergoing severe symptoms of withdrawal or, when the situation finally hits bottom, a shift in the entire world view (Learning III).

The problem is that the line between the two types of learning, acclimation and addiction, can prove to be somewhat blurry in the long run. What began as an ingenious adaptation can evolve toward pathology. The saber teeth of a tiger can have short-range survival value, but they vitiate flexibility in other situations that ultimately prove to be crucial. The rest of the system adapts so as to make the innovation less and less reversible; interaction with other species creates further innovations that push the situation towards runaway; flexibility is destroyed, and finally, the "favored" species is so "favored" that it destroys its own ecological niche, and disappears. In addiction "the innovator becomes hooked into the business of trying to hold constant some rate of change." What began as a gain at one level became a calamity in a larger context.

Human social systems provide many illustrations of this problem, and Bateson cites the history of DDT as a case in point. Discovered in 1939, the pesticide was deemed essential to increase crop yield and to save overseas troops from malaria. It was, Bateson says, "a symptomatic cure for troubles connected with the increase of population." By 1950, many scientists knew that DDT was toxic to many animals, but too many other variables had rearranged themselves to enable us to get "unhooked" from the pesticide. A vast industry had grown up around its manufacture; the insects at which the chemical was directed were becoming immune; the animals that fed on those insects were being exterminated; and in general, the use of DDT permitted an increase in world population. So now, we are addicted to its use, and nature is attempting a correction in ways that are frightening. DDT is now appearing in mother's milk; fish, if they do not become poisonous as carriers of mercury, may soon become so as carriers of DDT; forty-three species of malaria-bearing mosquitoes are now resistant to major insecticides, and the incidence of malaria in some countries has increased a hundredfold during the past fifteen years. What began as an ingenious ad hoc measure wound up exacerbating the original problem, eventually plunging us into an addictive spiral that now threatens our existence. [16]

For the time being, our reaction to this situation is to seek an increasingly larger "fix." Like the alcoholic we still believe that the answer lies in "rational mastery," and so escalate our insecticides to greater levels of toxicity, thereby making more dangerous insects immune, and so raising the battle to the next higher level. Perhaps when, as in some science fiction horror movie, giant mantises come knocking at the door, we shall finally comprehend that "rational mastery" was the problem; but by then it will be too late.

The so-called energy crisis is an equally cogent example of the addictive spiral. The columns of our newspapers are filled with articles that express concern over the coming disappearance of fossil fuels, and insist on the need to develop new sources of energy—especially nuclear energy—to meet the increasing demand. The voices suggesting that we might be "hooked" on energy, and that we had better move toward withdrawal rather than the next available "fix," have been largely drowned out by industrial interests that are committed to increasing the dosage of the "fix." Meanwhile, the negative feedback is becoming louder and louder, the near meltdown at the Three Mile Island nuclear plant in 1979 being only the most spectacular example. People living near freeway systems, according to one study done in Switzerland, are more likely to contract cancer than those farther away from high pollution density. Radioactive wastes are leaking out of containers buried deep in the ocean. Major blackouts occur in industrial areas, accompanied by widespread looting, while international conflicts over oil supplies and prices grow more intense. In short, the economy based on ever-expanding energy consumption is showing signs of severe strain. Modern industrial society is in effect trying to cheat the First Law of Thermodynamics, which says that it takes energy to deliver energy; that you never, in physics, get something for nothing. Using energy to solve the problems of industrial society is all of a piece with the mental framework of addiction. If Blake told us that energy was eternal delight, he also said that wisdom can be the result of pursuing folly to the limit. But once again, it may be too late. Our addiction may have brought the planet to the point of extinction.

Finally, the question of addiction can be applied to the whole style of Western life since 1600 A.D. To take an example from our earlier historical discussion, the Hermetic tradition was one of self-corrective feedback. Rational consciousness, especially in its emphasis on manipulating the environment, was kept in check (optimized), because it was simply one variable in a system organized around the idea of sacred harmony. With the advent of the Scientific Revolution came the attempt to maximize this particular variable. It was abstracted from its sacred context, and within a few generations what was once regarded as perverse came to be seen as normal. Unlimited expansion, ideologically ratified by the French Enlightenment and the economic theory of laissez-faire, began to make sense, and the need for an increasingly larger "fix" was regarded as part of the natural order of things rather than as aberrant. We are by now completely addicted to maximizing variables that are wrecking our own natural system. The emergence of holistic thought in our own time might itself be part of the general process of self-corrective feedback.

The preservation of diversity, point (4), which is crucial to the survival of all biological systems, is directly related to these problems, because it involves retaining flexibility rather than addictively consuming it.[17] Population geneticists have long been aware that the evolutionary unit is not homogeneous. Randomness, chance, is the source of anything new. Without diversity, there could be no emergence of new behaviors, genes, or organs for natural selection to operate upon. A wild population of any species has a wide variety of genetic constitutions spread throughout its individual members, and it is this heterogeneity that creates the potential for change essential for survival. Homogeneous situations, including the rigidity of addictive thinking, do not possess this resilience. Hence flexibility is itself part of the unit of survival, and of Mind. Love, wisdom, circuitry, optimization—all of these add up to an ethics of diversity, and it is this ethical system that Batesonian holism stands for. Yet all of Western industrial society, socialist or capitalist, officially strives for homogeneity, for unity of thought and behavior. In cities, Western man achieves single-species ecosystems, especially in architecture, design, and middle-class ideals of the "good life." In agriculture, he strives toward monoculture: fields upon fields of corn or soybeans, batteries of fowl producing eggs on the model of an assembly line. His ideas seem diverse, but they all ultimately stem from a Judeo-Christian tradition and the secular humanism of the Renaissance: the Golden Rule; survival of the fittest; premises of challenge (schismogenesis) and individual achievement; the nature of human "character traits" as fixed "entities," and so on. Some of these ideas may even be good (whatever that means), but having our heads filled with only one type of thinking cannot possibly be. Ultimately, this monomania is extended to everything and everyone we meet. As Lévi-Strauss wrote in *Tristes Tropiques*, Western secular humanism, in the name of respect for man, prescribes a single way of life and a single type of man. The joy of being with another person might be the aesthetic one of recognizing him or her as a human ecology different from oneself, manifesting the conscious/unconscious relationship in his or her own special way (each person is a song, as Gary Snyder has put it), but we typically hate the Other and demand that it be like us: safe, predictable, and in reality, a cliché.

And what is the truth, the ethics, that diversity speaks? It is, as Mary Catherine Bateson stated recently, and Nietzsche long before her, that we each have our own mythology, our own real possibilities to live out; that we are each "our own central metaphor." In the biological and ecological world, homogeneity spells rigidity and death. The natural world avoids monotypes because they tend toward weakness; they cannot produce anything new, and having little flexibility are easily destroyed. Systems that are reduced in complexity lose options, become unstable and vulnerable. Flexibility in personality types and world views provides, instead, possibilities for change, evolution, and real survival. Imperialism, whether economic, psychological, or personal (they tend to go together) seeks to wipe out native cultures, individual ways of life, and diverse ideas—eradicating them in order to substitute a global and homogeneous way of life. It sees variation as a threat. A holistic civilization, by contrast, would cherish variation, see it as a gift, a form of wealth or property.

Sometime ago, I had the pleasure of seeing a photography exhibit of European portraits from the 1920s and 1930s. The people in these pictures were "ordinary" people, not celebrities. What struck me most about the photographs was that it was absolutely clear that these were all distinct personalities, genuine individuals. One wanted to know them, for the eyes belied a sensation of complexity and idiosyncracy that might take years to unfold. I found the contrast between such faces, and the hollow, absent expressions of most contemporary urban dwellers, overwhelming. This same sort of organic diversity is celebrated by the American writer John Nichols in novels such as *The Milagro Beanfield War,* or by Fellini in a film like *Amarcord*, where almost everyone in the town has eccentricities that one might consider outrageous, but which, from another perspective, are quite splendid. Members of these communities fight endlessly over these differences, yet within the context of an instinctive understanding that they are all part of a larger ecology. The fighting becomes vicious only when the social ecosystem is threatened: in Nichols' case by capitalist notions of progress, in Fellini's by Fascism. If each character possesses (from our viewpoint) more than a slight touch of irrationality, the whole structure is itself rational, organic, whole. By contrast, in Western industrial societies each person is enjoined to fit a "rational," homogeneous, yet somehow "individualistic" (actually egotistic) stereotype, and the total effect is what both Bateson and Marcuse have described: senseless, crazy, a vast alienation rather than a vast ecology. It is the streamlining of life, whether in a Kansas wheatfield or in this year's graduating class at the University of Peking, which has, in its destruction of diversity, so impoverished human life.

9

The Politics of Consciousness

The sterility of the bourgeois world will end in suicide or a new form of creative participation. This is the "theme of our times," in Ortega y Gasset's phrase; it is the substance of our dreams and the meaning of our acts.

—Octavio Paz, *The Labyrinth of Solitude* (1961)

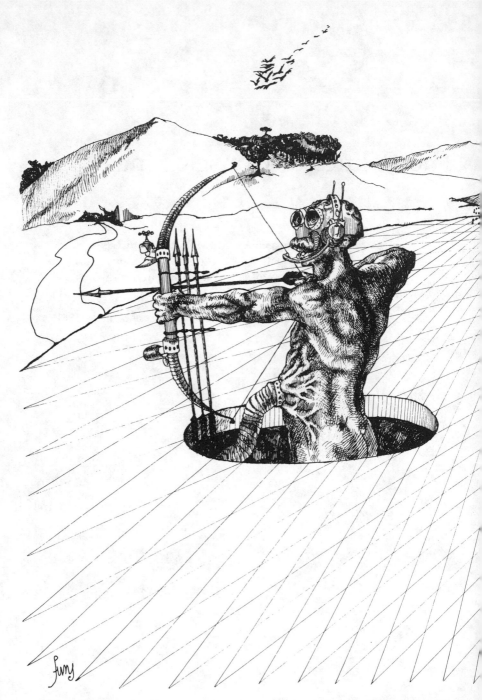

Plate 20. Fons van Woerkom, Illustration for Chapter 6 of Paul Shepard's *The Tender Carnivore and the Sacred Game* (1973).

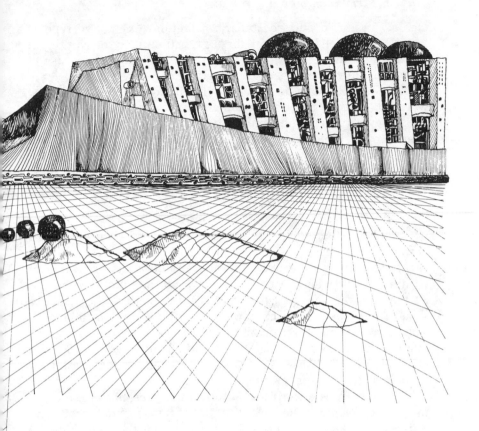

In 1883 or 1884, when my maternal grandfather turned five, he was sent by his parents to the *cheder*, or Jewish elementary school, where he would learn to read the Hebrew language and the Old Testament. It was the custom among the Jews of the province of Grodno (Grodno Guberniia) in Belorussia that each boy was given a slate upon entry to the *cheder*. It was his personal possession, on which he would learn to read and write. And on that first day, the teacher did something quite remarkable: he took the slate, and smeared the first two letters of the Hebrew alphabet—*aleph* and *beys*—on it in honey. As my grandfather ate the letters off the slate, he learned a message that was to remain with him all his life: knowledge is sweet.

And yet, the message is far more complex than this, for the act is almost an anthropological ritual with a rather layered symbolism. At the obvious level, the slate will be used for learning discursive

Hebrew grammar and vocabulary, a literal, nonemotive type of knowledge which is necessary for our functioning in the world. But the fact that the letters are tasted evokes an older, poetic use of language which is especially characteristic of Hebrew: the power of the Word. Hebrew is an unusually onomatopoeic language. The words often come close to creating an emotional resonance with what they represent conceptually. One of the messages being delivered in this honey-tasting ceremony is that real knowledge is not merely discursive or literal; it is also, if not first and foremost, sensuous. In fact, it is very nearly erotic, derived from bodily participation in the learning act. *De gustibus non est disputandum*, goes a Scholastic saying; about things eaten, there can be no argument. Or as the Sufis put it, those who taste, know.

There is, furthermore, a deliberate fusion here, even *confusion*, between discursive and sensual modes of knowing. As we have seen, identification (*mimesis*) and discrimination are both present within the physiological response system of the human organism. At the very moment that the child is introduced to the symbolic system that makes abstract thought, and thus categorization, possible, he performs the primal act of identification, the act of the infant, who puts everything in its mouth. Thus union and separation, self and other, are irrevocably intertwined in this first formal acquaintance with the learning experience.

Finally, there is a third level of meaning present here, one reminiscent of some of the insights of Lévi-Strauss. What is real here is ingested, taken into oneself. The symbolism is that of making the unfamiliar familiar: we literally eat the other, take it into our guts, and as a result are changed by it.

The recognition of these two last levels of knowledge is almost wholly absent from the institutions of official culture and education in contemporary Western society, steeped as they are in scientism and purely discursive knowing. Indeed, it is an immense irony that the "information explosion" of the modern era actually represents a *contraction* of our knowledge of the world, as the quote from Octavio Paz, in the epigraph to this book, clearly points out. Bateson, Reich, Jung, and a very few others represent the healthiest possible response to this state of affairs: the attempt to fight our way out of the cognitive corner into which we have painted ourselves. Theirs, as Theodore Roszak once remarked, is the search for live options, not the pursuit of moribund research which typically characterizes the "advanced" thinking of our modern university system. Digital knowledge is not necessarily wrong in itself, but pathetically incomplete, and thus it winds up projecting a fraudulent reality. University personnel, and more broadly the techno-bureaucratic elite of Western culture, are paid pretty much in proportion to their ability to promote and maintain this world view. In this way, analogue reality is suppressed, confined, or at least domesticated.

Yet the whole situation is unstable for reasons already indicated. Not merely does our analogue side fight back, but purely digital knowledge, since it is never "ingested," never "sticks to our ribs." The whole situation is a charade, because no real emotional commitment beyond economic payoff and ego-gratification is involved. We have been bewitched into believing that these rewards are fundamental, but a deeper, nagging voice tells us otherwise. Indeed, the danger of such a bloodless type of knowledge, and of the fact-value distinction in general, was not lost on one of its greatest defenders, Max Weber, in his classic essay, *The Protestant Ethic and the Spirit of Capitalism:* "Specialists without spirit, sensualists without heart; this nullity imagines that it has attained a level of civilization never before achieved."[1]

It was my grandfather's fortune to be born and raised in a world in which the sacred and secular were still closely united. In the cloistered community of the Russian *shtetl,* he never had to face the dilemma recognized by Weber. But it was also his fate to leave the *shtetl,* to emigrate first to England and then America, and thereby be exposed to the secular tide of the modern world. For the rest of his life, he was condemned to struggle with the great metaphysical problem of our age: how to reconcile what he knew in his head, with what he knew in his heart. Very obviously, I inherited this struggle, and this book represents at least a part of my attempt at a resolution.

What do I know in my heart, then? I know that in some relational sense, everything is alive; that noncognitive knowing, whether from dreams, art, the body, or outright insanity, is indeed knowing; that societies, like human beings, are organic, and the attempt to engineer either is destructive; and finally, that we are living on a dying planet, and that without some radical shift in our politics and consciousness, our children's generation is probably going to witness the planet's last days.

I also know some important things in my head. I know that the occult revival of our times is a response to these events, and in

269

general I believe that the archaic tradition, including dialectical reason and various psychic abilities that all of us possess, are important things to revive. But for the most part, I see our immediate future in a post-Cartesian paradigm, not in a premodern one. I know that despite its abuse, intellectual analysis is a very important tool for the human race to have, and that ego-consciousness is not without its survival value. And I know that any meaningful resolution of the fact-value distinction must go beyond one's own personal individuation; it must be social, political, environmental. When Sartre wrote that man is condemned to be free, he meant not this or that man (or woman), but the whole human race.

My thesis about Bateson is that in terms of resolving these difficulties, and getting the sacred and the secular back together again, his work represents the best we have up to this point. This is not to say that his holistic paradigm is problem-free, and I shall explicate some of these problems later on in this chapter; but its chief advantage is that it embraces value without sacrificing fact. It is a mature type of alchemical/dialectical reasoning adapted to the modern age. I have spent some time demonstrating its superiority to the Cartesian paradigm, and suggesting its formal similarities to the Hermetic world view and traditional systems of thought. I have argued that in Bateson's work, Mind is abstracted from its traditionally religious context and shown to be a concrete, active scientific element (process) in the real world; and that in this way, participation exists, but not in its original, animistic sense. Before moving on to a critique of that work, then, I wish to summarize what I regard as the unique triumphs of the Batesonian paradigm, in particular its superiority to the archaic tradition with which it nevertheless has much in common.

The chief advantage of Batesonian holism over the archaic tradition is its self-conscious character. Mind, as I have noted, is present in the latter, but in an undifferentiated sense ("God"). Bateson's conception of Mind is specific; he is able to delineate its characteristics in an explicit way. Thus he is not advocating a direct revival of archaic knowledge, but a type of self-conscious *mimesis* in which we would soften and work with the conscious/unconscious dichotomy rather than simply attempt to dissolve it. Emotion has precise algorithms, and in his studies of the analogue and relational nature of reality Bateson has given us clear examples of how this reality can be charted. The differences between archaic thought, modern science, and Batesonian holism can be seen in Chart 3. The pure materialism of modern science stands out starkly

Chart 3. Comparison of schizophrenia in three world views

	Archaic tradition	Cartesian paradigm	Batesonian holism
Interpretation	Possession by spirits	Organic disease (genetics, brain chemistry, etc.)	Deutero-learning (in the family) into a pattern that masks the nature of metacommunication (the double bind)
Treatment	Exorcism (purely spiritual)	Alteration of the molecular operation of the brain with drugs or shock (purely mechanical)	Work on the schizophrenic system through family therapy, so that person starts to metacommunicate properly. Therapist takes role of pointing out double bind, so that it can be broken.
Results	Probably mixed. Resolution is individual, personal, internal.	Effective in suppressing symptoms. Soul or spirit crushed; person becomes a "productive member of society." Resolution individual, but externally imposed.	Too early to tell, beyond Laing's work and that of a few others. Effectiveness depends on disrupting the schizophrenic system, i.e., revealing the organized pathology of the family. These are internal changes with radical social implications.
Type of society implied	Spiritual/religious	Scientific/materialistic, organized around the notion of productivity and efficiency. Logical end point: a reified, uniform, dystopian nightmare.	Self-realizing: one immersed in primary process and analogue communication. Extended family system with awareness of wide relational reality and the importance of healthy metacommunication. Goal of this society neither God (salvation) nor achievement, but healthy relations.

271

here, whereas the nonmaterialism of the first and third columns causes them to exhibit a formal similarity. For example, consider the schizophrenic who constantly talks to himself in conflicting, hallucinating voices.[2] The approach of Western medicine fails to recognize what both the theory of possession and the theory of the double bind know: that this individual got caught up in an alien Mind, or mental system; that this Mind or system has literally invaded him; and finally, that it is fully real. A person caught in a schizophrenic double bind, as we have seen, cannot speak his own mind, for he has learned that there are severe penalities for doing so. In this sense, the boy put on display by Kraepelin was indeed possessed by an alien spirit, and had he lived in the Middle Ages it is very possible that exorcism would have driven it out. Yet such an explanation is not possible in a scientific age, and this is where Bateson's approach is so valuable. If we can accept the notion of consciousness as being fully real, and understand how it got shaped into a certain type of Mind (mental system) so as to include the boy and his family *and* their way of relating to him, we are then in a position to break that double bind and create a different, and healthier, Mind. Furthermore, such analysis and resolution is not confined to single individuals, as is the archaic or scientific approach. As is so clearly the case in Laing's work, the entire family structure is implicated, along with the society that is made up of such neurotic (and psychotic) building blocks. Though exorcism is probably superior to thorazine, and certainly more humane, neither means is concerned with the political conditions that produced the craziness in the first place. Batesonian analysis does not go as far as it could in this regard, but it is an important start.

Similarly, the archaic tradition understood certain things about light and color (Goethe being its last modern representative), or electricity and gravity, that modern science has left out; but it is no longer possible for us to see these phenomena in teleological terms, or as direct manifestations of God or a life-force. Nor would a purely spiritual interpretation open up any fruitful line of inquiry in such cases.[3] But as I suggested in Chapter 6, analysis of these phenomena which proceeds in terms of a "detached observer" is also obsolete. Batesonian holism, on the other hand, could offer a nonspiritualist, process-oriented mode of investigation. One could see such phenomena cybernetically, or systemically, as part of a Mind that includes the investigator (including his or her affective responses) in it. A Batesonian analysis would study not just the quantitative relations but the qualitative ones: the essential arrangement present, the levels of Mind and the nature of their interaction.

It should also be noted that the essence of cybernetic explanation itself, the insistence on the relational nature of reality, which is absent from the Cartesian paradigm, is also present in the archaic tradition. Traditional cultures had an intuitive grasp of the cybernetic concept of circuitry through practices such as totemism and nature worship, and in this way managed to preserve and protect their environment. By explicating the interrelations between the sub-Minds around us on a Batesonian model, we could learn not to pollute Lake Erie because the resulting chain reaction would be immediately evident to us. The advantage here is sane, holistic behavior without a return to complete *mimesis*. In a Batesonian framework, as opposed to archaic consciousness, we can actually focus on the circuit, not just be immersed in it. The hope is that archaic knowledge, especially the recognition of Mind, will emerge under an aesthetic rubric, so that our science (knowledge of the world) will become artful (artistic). The hope is that we can have both *mimesis* and analysis, that the two will reinforce each other rather than generate a "two cultures" split. Only through a mimetic relationship with your environment (or anything you address, for that matter), can you obtain the insight into reality which will then form the center of your analytical understanding. Fact and value merge and Mind is revealed as both a value and a mode of analysis.

Finally, Bateson's concept of Learning III, the psychological breakthrough to a "vast ecology," is nearly identical to the religious conversion of the archaic tradition, whether in Christian mysticism, the Zen *satori*, or the final stage of alchemical transmutation. Bateson does not explicitly advocate any of these practices, yet it is clear that in Learning III, as in these traditions, the central event is a redefinition of personality. One breaks through to a new level and gets a perspective on his or her own character and world view. There is, however, an important difference between Bateson's notion of Learning III and traditional self-realization: Bateson's concept is an integral aspect of the search for community and fraternity, not (as in Norman O. Brown, for example) merely a personal ecstatic vision. In Bateson's study of Alcoholics Anonymous, the Higher Power to which the alcoholic finally surrenders is not only "God" (or the unconscious), but the other members of

AA. He makes himself a part of their social reality, their common struggle. Thus no matter how or where you discover Mind, says Bateson, "it is still immanent in the total interconnected social system and planetary ecology."[4]

I wish now to turn to a critique of Bateson's work, but must first share with the reader a quandary I have about doing so. In attempting to draw up a critique, I quickly discovered that it was not possible to construct one in an abstract, conceptual way. The critique rapidly became political, and perhaps this is not surprising. Historically, politics and epistemology have had an uncanny way of reinforcing each other; and in the case of Bateson's work, the union of fact and value is so close that to explicate epistemology is necessarily to explicate ethics, and thus unavoidably, politics. As I am sure the reader understands, much of my interest in Bateson stems from the hope of finding a liberatory epistemology; which also means, as far as I am concerned, a liberatory politics. Although liberation is clearly implied in the Batesonian paradigm, its formal similarities to the dialectical tradition make it liable to the type of political ambiguity which has bedeviled this tradition historically. One gets a left-wing Reich and a right-wing Jung; the revolutionary religious cults described by Christoper Hill,[5] and the authoritarian self-realization groups (*est*, the "Moonies," the Church of Scientology) which currently plague the American scene. Although Bateson personally had no truck with right-wing politics, a number of his concepts are double-edged; they have the potential for oppression as well as liberation. Political ambiguity and epistemological ambiguity go hand in hand here, and it is this ambiguity that is the focus of my critique. Before the critique can be made with any clarity, then, it will be necessary to sketch out the liberatory political vision that is consonant with the Batesonian paradigm.[6]

One of the most obvious characteristics of a future "planetary culture" will be the straightforward revival and elaboration of analogue modes of expression, a process that will involve the deliberate cultivation and preservation of (digital) incompleteness. Such a culture will be dreamier and more sensual than ours. The inner psychic landscape of dreams, body language, art, dance, fantasy, and myth will play a large part in our attempt to understand and live in the world. These activities will now be seen as legitimate, and ultimately crucial, forms of knowledge, and will be accompanied by a direct cultivation of psychic faculties: ESP,

274

psychometry and psychokinesis, aura reading and healing, and others.[7] Simultaneously, there will be a strong shift in medical practice toward popular and natural healing; an avoidance of drugs and chemical manipulation; and a near merger with ecology and psychology, since it will be widely recognized that most disease is a response to a disturbed physical and emotional environment. Birth will not take place on the "assembly line" of the modern hospital, but at home, so that the gentle birth practices described in Chapter 6 can once again shape childhood development.[8] In general, the body will be seen as part of culture, not a dangerous libido to be kept in check, a change in perception which will involve a drastic reduction in sexual repression, and a greater awareness of ourselves as animals. This future culture will also see a revival of the extended family, as opposed to the competitive and isolating nuclear family that is today a seedbed of neurosis. The elderly will be mixed in with the very young, rather than dumped in old-age homes for the "unproductive," and their wisdom will be a continuing part of cultural life.

Such changes will enable a parallel shift in the ideal of personality, specifically a shift in focus from the ego to the self, and they will encourage the interaction of this self with other selves. The result will be an emphasis on community rather than competition, on individuation rather than individualism, and an end to the "false-self system" and role-playing that have so badly desecrated (desacralized) human relationships. As for power, it will be the equivalent to centeredness, inner authority, and not the ability to make others do what you want them to against their will. Power will be defined as the ability to influence others *without* pressure or coercion; the phrase "position of power" will be recognized as a contradiction in terms, for it will generally be understood that if a person needs a position to feel his or her power, then what he or she is really feeling is impotence.[9]

The future culture will have a greater tolerance for the strange, the nonhuman, for diversity of all sorts, both within the personality and without. This increase in tolerance implies a shift from the Freudian-Platonic to the alchemical notion of sanity: the ideal will be the "many-aspect" person of kaleidoscopic traits, who has a greater fluidity of interests, working and living arrangements, sexual and social roles, and so on. All behavior will be seen as having at least one complement, or "shadow," in need of legitimate expression. There may also be experimentation with modes of thought and relationship which are nonschismogenic—an attempt to create behavior patterns that are not cumulative and which are inherently satisfying rather than dependent upon delayed gratification.[10] The principle of diversity will require the preservation of endangered species and endangered cultures, as factors that enlarge the gene pool of possibilities and thereby make life more stable, durable, and interesting.

Human culture will come to be seen more as a category of natural history, "a semipermeable membrane between man and nature."[11] Such a society will be preoccupied with fitting into nature rather than attempting to master it. The goal will be "not to *rule* a domain, but to *release* it"; to have, once again, "clean air, clean clear-running rivers, the presence of Pelican and Osprey and Gray Whale in our lives; salmon and trout in our streams; unmuddied language and good dreams."[12] Technology will no longer pervade our consciousness and its presence will be more in the form of crafts and tools, things that lie *within* our control rather than the reverse.[13] We will no longer depend on the technological fix, whether in medicine, agriculture, or anything else, but instead favor solutions that are long-term and address themselves to causes rather than symptoms.

Politically, there will be a tremendous emphasis on decentralization, which will extend to all the institutions of society and be recognized as a prerequisite to planetary culture itself. Decentralization implies that institutions are small-scale and subject to local control, and that political structures are regional and autonomous. Characteristic of such decentralization are community hospitals and food cooperatives, the cultivation of neighborhood spirit and autonomy, and the elimination of such destroyers of community as television, automobiles, and expressways. Mass production will yield to craftsmanship, agribusiness to small, organic, labor-intensive farming, and centralized energy sources—especially nuclear power plants—to renewable energy options appropriate to their own regions. Mass education centers teaching essentially one type of knowledge as preparation for a career will be replaced by direct apprenticeship, in the form of a lifelong education that follows one's changing interests. One will not have a career, but a *life*. The blight of suburbs and urban sprawl, truly the antithesis of city life, will be replaced by a genuine city culture, one native to its own region rather than reflecting an international world of mass communication. The city will once again become a center of life and pleasure, an *agora* (that fine Greek word), a market place and meeting place, Philippe Ariès' "medley of colors." People will live closer to their work, and in general there will not be much distinction between work, life, and leisure.[14]

The economy, finally, will be steady state, a mixture of small-scale socialism, capitalism, and direct barter. This will be a "conserver" society, with nothing wasted and a great emphasis, to the extent that it is possible, on regional self-sufficiency. There will be little interest in profit as an end in itself. The posture toward others, and toward natural resources, will be one of harmony rather than of exploitation or acquisition. As ecologists Peter Berg and Raymond Dasmann have written, economics will become "ecologics," a subbranch of ecology.[15]

How are we going to get there? From the present vantage point, the vision of a future in which fact and value are once again reunited, in which men and women have control over their own destinies, and in which ego-consciousness is more reasonably situated within a larger context of Mind, seems utopian in the extreme. Yet as Octavio Paz observed, the only alternative is suicide. Western industrial society has reached the limits of its own deutero-learning, and much of it now is in the midst of the social analogue of either madness or creativity, that is, re-creation (Learning III). Given this situation, how utopian is such a vision? Of course, if one believes that only violent revolution produces substantive change, and that such a transformation can be accomplished within a few decades, then planetary culture does not have much of a chance. If, however, we are talking of a change on the scale of the disintegration of the Roman Empire, such has been suggested by Theodore Roszak, Willis Harman, and Robert Heilbroner, among others, then our utopian vision starts to appear increasingly realistic.[16] In fact, one of the most effective agents of this set of changes is the decay of advanced industrial society itself. Thus Percival Goodman writes in *The Double E* that the conserver society will not come about because of voluntary effort, but because the planet simply cannot support the world of an ever-expanding Gross National Product. Industrial economies are starting to contract. We may choose to make a virtue out of what has been called "Buddhist economics," but we shall have to return to a steady-state economy whether we like it or not.[17]

Social change is also being generated by millions of individuals who have no interest in change per se, but have effectively undertaken an "inner migration," or withdrawal. Both Harman and Heilbroner have pointed to the fact that the industrial economies are going to face a severe economic crunch at the very time that their workers, both blue and white collar, have found their work so devoid of intrinsic value that they are increasingly finding meaning elsewhere, and privately withdrawing their allegiance from their jobs. The Protestant work ethic, the spiritual support of our present way of life, will not be there when the economy needs it most. A 1975 report of the Trend Analysis Program of the American Institute of Life Insurance predicts a weakening of "industrial era philosophy" during the next two decades, with concomitant worker alienation, slowdowns, sabotage, and riots. "We may," concludes the report, "be somewhere in the middle of a turbulent transition to a new, or at least somewhat different culture," beginning about 1990.[18]

On the political level, decay will probably take the form of the breakup of the nation-state in favor of small, regional units. This trend, sometimes called political separatism, devolution, or balkanization, is by now quite widespread in all industrial societies. The number of new nations has risen dramatically since 1945, and other societies are beginning to fragment into provincial and sectarian subunits. Leopold Kohr predicted this trend (enthusiastically) as early as 1957 in his book, *The Breakdown of Nations;* official culture, such as *Harper's,* is now terrified of it. More soberly, a group of about 200 European experts, in the book *Europe 2000,* sees the revolt of a regional periphery as very likely.[19] There are now strong separatist movements not only in the United States (Northern California, Upper Michigan, Idaho's Panhandle), but in Scotland, Brittany, Pays Basque, and Corsica; and many other countries are also experiencing strong regional sentiments, so much so that the Europe of 2000 A.D. may well look like a mosaic of very small states. This process represents a reversion to original political boundaries that existed prior to the rise of modern nation-states: not France, but Burgundy, Picardy, Normandy, Alsace, and Lorraine; not Germany, but Bavaria, Baden, Hesse, Hanover; not Spain, but Valencia, Aragon, Catalonia, Castile; and so on. In general, writes Peter Hall, what at all levels

> used to be called separatism and is now usually called regionalism— fundamentally the desire and willingness to assume more direct control over one's own destiny—is perhaps the strongest political drive

now operating: it is the main cause of the "crisis of authority" and the weakening of centralized control. [20]

Holistic society is thus coming upon us from a variety of sources that cut across the traditional left-right political axis. Feminism, ecology, ethnicity, and transcendentalism (religious renewal), which ostensibly have nothing in common politically, may be converging toward a common goal. These holistic movements do not represent a single social class, nor can they even be analyzed in such terms, for by and large they represent the repressed "shadows" of industrial civilization: the feminine, the wilderness, the child, the body, the creative mind and heart, the occult, and the peoples of the nonurban, regional peripheries of Europe and North America—regions that have never bought into the ethos of the industrial heartland and never will. If there is any bond among the elements of this "counterculture," it is the notion of recovery. Their goal is the recovery of our bodies, our health, our sexuality, our natural environment, our archaic traditions, our unconscious mind, our rootedness in the land, our sense of community and connectedness to one another. What they advocate is not merely a program of "no growth" or industrial slowdown, but the direct attempt to get back from the past what we lost during the last four centuries; to go backward in order to go forward. In a word, they represent the attempt to recover our future.

What is remarkable in many of these developments, also, is the attempt to create a politics that does not substitute one set of rulers for another, or even one political structure for another, but which reflects the basic needs of mind, body, sexuality, community, and the like. The goal, notes that ancient Chinese oracle, the *I Ching*, is

a satisfactory political or social organization of mankind. [Therefore] we must go down to the very foundations of life. For any merely superficial ordering of life that leaves its deepest needs unsatisfied is as ineffectual as if no attempt at order had ever been made. [21]

In various ways, this has become the goal of all holistic politics; a politics that would be the end of politics, at least as we know it today.

If all of these changes, or even a third of them, came to pass, the anomie of the modern era would surely be a closed chapter in our history. Such a planetary culture would of necessity erase our con-

279

temporary feeling of homelessness, and the sense that our personal reality is at odds with official reality. The infinite spaces whose silence terrified Pascal may appear to men and women of the future as extensions of a biosphere that is nurturing and benevolent. Meaning will no longer be something that must be found and imposed on an absurd universe; it will be given, and, as a result, men and women will have a feeling of cosmic connectedness, of belonging to a larger pattern. Surely, such a world represents salvation, but only in the sense that there is no need to be saved in the first place. A loss of interest in the traditional opiates would likely follow, and even psychoanalysis would be seen as superfluous. What would be worshipped, if anything, is ourselves, each other, and *this earth*—our *home*, the body of us all that makes our lives possible.

This, then, is the liberatory version of a planetary politics that is congruent with the epistemology of Batesonian holism. It is my hope that the social and political developments of the next century move us closer to such a world. However, as indicated earlier, things are not that simple, because a number of Bateson's concepts are double-edged. I do not mean to suggest that consciousness by itself makes history (there *is* no consciousness by itself!), but that the two form a gestalt, and that Batesonian holism is potentially congruent with political configurations less benevolent than the one outlined above. In fact, should political developments make ideological use of holistic concepts, and wind up emphasizing certain aspects of these as opposed to others, we could be victimized by a rather grim twist: the specter of holistic consciousness as the agent of even more alienation, more reification, than we have at present. This possibility merits further investigation.

The original context of Batesonian holism was hardly (in Theodore Roszak's phrase) the "Taoist anarchy" sketched above, but the rigid hierarchical society of the British aristocracy. We have seen that most of Bateson's scientific concepts were adumbrated in the work of his father; and in his exposition of William Bateson's work, William Coleman correctly identifies the ingrained political conservatism that characterized the context of that work.[22] The England of the late nineteenth century was in the grip of a profound pessimism: a disenchantment with utilitarianism, democracy, and parliamentary politics. The glittering promise of Crystal Palace (1851) had not materialized, and the pervading mood was one of civilization in collapse. The intelligentsia and the upper classes reacted by returning to traditional values, notably aesthetic sensibility, intuitionism, and an organic conception of society. These three traditional conservative themes, says Coleman, were central to William Bateson's thought. His emphasis was on the genius, the exceptional person, whose development would never be encouraged in an egalitarian society. William Bateson's interest was in vision and inspiration, not in ambition and calculating reason, hence his revealing remark at the end of the Great War: "We may have made the world safe for democracy, but we have made it unsafe for anything else." As Coleman notes, he saw the world of commerce and democracy as a veritable dark age. For the elder Bateson, the natural hierarchy of function in the biological world validated class society, and he held that correct political solutions were those that managed to *preserve* inequality, to coordinate the different and unequal parts of society in the performance of their proper job.

Given this extreme elitism, many of William Bateson's scientific concepts take on a peculiar light. The primacy of form and pattern (Mind) over matter reflects a mentality that pits the lofty *Geist* of aristocratic intellectualism against the grubby materialism of middle-class commerce and professionalism. The notion that variation comes from within, rather than from the external action of the environment, may certainly have a long alchemical ancestry (as we saw in the case of Newton), but in William Bateson it reflected the aesthetic sensibility of inner purity and intuitionism: the lotus in the cesspool, the man above the crowd. A similar type of class consciousness characterized his defense of the classics, and the notion of true education as an "awakening to ecstasy"—a view that assumes that most people are trapped in Plato's cave. Perhaps most revealing is William Bateson's central holistic principle, that any variation must result in a coordinate change in the entire organism being affected. In 1888 he wrote his sister that unless such correlated variation occurred, a system could not continue to be a system. Stated in this way, Bateson's principle has strong political overtones; it reflects a bias against change per se and especially against any form of disturbance. As one who had succeeded in entering elite circles, William Bateson did not want the system that had nurtured him to disintegrate. In his science, as in his politics, the maintenance of stability became the core of reality, and any but the most gradual and organic changes were to be viewed with deep suspicion and hostility—an outlook that put him squarely in the

281

tradition of Edmund Burke. Since Gregory's own scientific concepts were so strongly shaped by those of his father, we should not be surprised to find that they have—or can have—political implications that echo this extreme conservatism. In what follows I wish to focus on the following concepts or aspects of Gregory's work: the emphasis on communication and information exchange, the Theory of Logical Types, homeostasis, and Learning III.

As we have seen, the transmission of ideas around a circuit is central to cybernetic explanation. It makes possible the refutation of Cartesian atomism and mechanical causality in favor of something called Mind and its interrelations with other Minds. We have also seen how superior the latter is to the former in dealing with schizophrenia, alcoholism, learning theory, and other areas of research. The problem arises when the notion of information exchange is applied to situations that are blatantly and immediately political.[23] Anthony Wilden gives the following example:[24]

Person A: Please give me a glass of water.
Person B: (Hands water to A)
Person A: Thank you.

We can, of course, analyze the interchange as an exchange of messages, and at face value it would seem that A is the supplicant, submissive to B, or that they are perhaps equals. However, says Wilden, suppose the reality of the situation is that A's request was in fact a command? Suppose A is a man, and B a woman? Suppose A is a foreman, and B a factory worker or a sharecropper? Suppose B is black, or on welfare? What is truly operative then can only be found in an analysis of the history of race, or sexuality, or vested interests. It cannot be found in an analysis of messages alone, or of disturbed communication. Schismogenesis may serve to explain the nuclear arms race or domestic strife, but in general it is doubtful that war is a failure of communication, and I suspect that the North Vietnamese knew perfectly well what the Americans were up to. The same can be said of the so-called generation gap of the 1960s, in which the media were able to avoid taking student opposition to the dominant culture seriously by turning it into a "communications" problem. Explanation at this level deals only with the here and now, with what is manifest, and it presupposes a society of equals, an open or pluralistic situation in which all conflicts are capable of smooth resolution once the blocked channels of communication are cleared. Used in this way, cybernetic theory is not a form of liberation but of mystification. The relationship of oppressor to oppressed is not typically a problem of semantics,[25] and such an emphasis can easily serve to reinforce that relationship, though such was certainly not Bateson's intent.

The Theory of Logical Types, employed so brilliantly by Bateson, shares a similar political bias.[26] In essence, it is a theory of hierarchical relationships, and it is conceivable that a logic of classes implies a class society, or at least one in which some groups have a higher social or theoretical status than others. Logical typing reflects and implies a top-down attitude toward power, although this attitude is muted in the social analysis based on the Theory of Logical Types. This political bias, however, was not lost on one of the coauthors of the theory, Bertrand Russell, who remarked at one point in his *Autobiography* that he saw the theory, at the time of its formulation, as a contribution to the preservation of British hegemony and world order. Although logical typing is obviously a powerful tool for understanding certain phenomena, it is not clear that it has a very wide application; yet it is absolutely central to cybernetic analysis, as Bateson would be the first to admit.

As it turns out, Russell admitted his doubts about the theory to Cambridge mathematician G. Spencer Brown in an exchange that occurred in 1967. Brown had developed a mathematical proof that demonstrated that the theory was unnecessary, and showed it to Russell. Russell agreed, adding that it was "the most arbitrary thing that he and Whitehead had ever had to do, not really a theory but a stopgap. . . ."[27] An indirect refutation of logical typing, moreover, was developed in 1945 by the cybernetic theoretician Warren McCulloch, who argued for a *heterarchy* of values rather than a hierarchy. By means of a mathematical analysis of the central nervous system, McCulloch showed that values were not magnitudes and thus that transitivity (inequality of relationships) could not be applied to them.[28] One can, for example, establish a hierarchy or wavelength of frequency for the colors of the spectrum, but there is no way to prove that red is somehow "better" than blue, or the reverse. But McCulloch never developed his analysis further, probably because cybernetic theory would have been seriously attenuated if logical typing were invalidated. The fact remains that heterarchy implies egalitarianism, and hierarchy, a world of classes and orders. But there is no way one can demonstrate that hierarchy is validated by the natural world.[29]

Third, we have the concept of homeostasis, with its obvious roots in William Bateson's principle of correlated variation, and again the conservative implications are obvious. As René Dubos was quick to point out, taken to its logical conclusion, homeostasis says that "whatever is, is right." Dubos thus argues for "homeo-kinesis," or what C. H. Waddington calls "homeorhesis": "sta-bilized flow rather than stabilized state."[30] Politically, the con-cept of homeostasis leads logically to quietism, to passivity in the face of an oppression that is seen to be "in the order of things" (otherwise it wouldn't have happened!). Bateson's point, of course, is that interference frequently makes things worse, and that revolution is often just that—a revolving door, a change of masters rather than a change of values. It is an important point, but it is hardly true that all fight for freedom is futile. Nor does Bate-son's approach come to terms with the totalitarianism that might emerge if the powers that be were, due to any lack of resistance, given free rein.

As in the case of information exchange, the issue may be how and where the concept is applied. Early cybernetic writers used closed systems, such as the thermostat, as their paradigm. A ther-mostat may be "alive" in some cybernetic sense, but it is closed in that it does not exchange any material with its environment, and its final state is determined by its initial conditions. Open systems (a forest, a nation) do exchange material with their surroundings, and their final states are not predetermined. As a result, they are open to substantive change (whether it occurs or not). In other words, only closed systems are truly homeostatic, returning always to their original starting point. Homeostasis is thus only a special case of open systems.[31] The latter can undergo homeorhesis, change that is part of the overall developmental program (language acquis-ition, puberty), or "morphogenesis," change that proves to be an alteration of the program itself (Learning III, the Scientific Revolu-tion, the collapse of the Roman Empire—all of which can be "pre-dicted" only in retrospect).[32] Bateson is fully aware of the dif-ference between open and closed systems, but his overriding em-phasis is on stability rather than alteration; for example, how sym-metrical schismogenic situations manage to trigger their comple-ment so as to mitigate the threat of disintegration, or how an ecosystem struggles to maintain itself by generating negative feed-back. Bateson does say that the process of maintenance will not necessarily bring the system back to its initial starting point, but his general emphasis on the maintenance of internal consistency tends

284

to put change in the category of an undesirable event. Thus he likens change to a tear, a rent in the fabric of things, and the process of maintenance to healing or mending.[33]

Such an emphasis on homeostasis and stability, of course, can certainly be seen to be congruent with the small-scale, ecological, decentralized "conserver" society described above. But on a strictly homeostatic model, we would never get there, whereas the likelihood is that we are in the midst of a vast and violent morphogenesis. Furthermore, the cybernetic model of society is not congruent only with the conserver society, as several critics have pointed out. It can easily be used to validate the alternative model of industrial totalitarianism. There is, for example, nothing intrinsic in Bateson's work that implies decentralization. The cybernetic model could well describe a mass society managed by social engineers through a series of "holistic," bureaucratic parameters, and indeed, precisely this scenario is envisioned by Robert Lilienfeld in his book *The Rise of Systems Theory*. Far from leading to a planetary culture, says Lilienfeld, the emphasis on communications suggests a world knit closely together by a system of computerized mass media and information exchange.[34] Such a world would be the *end* of diversity and freedom, a homogenization of the globe under man's dominion—or rather, under the dominion of a small, powerful elite. One thinks here of Interpol, or the data banks that continue to be assembled on the citizens of industrial societies, soon to be transferred to silicon chips, microcomputers that could easily be made available to the police, the government, and even to hospitals and banks. "Systems science," wrote one of its founders, Ludwig von Bertalanffy, "centered in computer technology, cybernetics, automation and systems engineering, appears to make the systems idea another—and indeed the ultimate—technique to shape man and society even more into the 'megamachine.'"[35] Bureaucracy and centralization could become the order of the day, in which the concept of hierarchy, or logical typing, would mean that the lower ranks were "free" to obey the upper ones, to fall into homeostatic step with them. This situation, with its obvious echoes of *Brave New World* or *1984*, is hardly the vision of holistic harmony Bateson had in mind, but it is as much implied by his epistemology as the utopian scenario previously outlined, and the concepts of information exchange and the rest could be used to rationalize it.[36]

Part of the problem, perhaps, is that neither cybernetics nor ecology is immune to mechanistic treatment. As Carolyn Merchant has pointed out in *The Death of Nature*, the dominant trend in American ecology studies since the 1950s has been reductionist and managerial. On this model, she notes, data

> are abstracted from the organic context in the form of information bits and then manipulated according to a set of differential equations, allowing the prediction of ecological change and the rational management of the ecosystem and its resources as a whole.

The word "ecosystem," in fact, was developed by this school of thought to replace the more anthropocentric and decentralized phrase, "biotic community." The approach here is globalist, and computer-based reports, such as the Club of Rome's famous *Limits to Growth* (1972), which make recommendations for managing the resources of the entire world, are the logical descendants of this branch of ecology. As Merchant points out, the same criticism can be made of much of systems theory. Its proponents often claim that their approach is holistic, but a gestalt is an intangible thing. The chances are that once mathematized, it stops being a true gestalt.[37]

Cybernetic thinking, in short, does not automatically take us out of the world of Francis Bacon. The cybernetic mechanism may be a more sophisticated model than the clockwork model of the seventeenth century, but it is still, in the last analysis, a mechanism. Bateson's experiment with the dolphin, for example—driving it crazy until a clear-cut result was obtained—is as good an example of Bacon's *natura vexata* as any.[38]

Finally, we come to the issue of Learning III, the "awakening to ecstasy," or sense of merger with a "vast ecology." As noted above, Bateson does not explicitly advocate meditation, yoga, alchemy, or whatever; his is a self-conscious *mimesis* that does not dispense with cognitive knowing. But in lieu of such practices, how is the insight or breakthrough of Learning III to be achieved? The alcoholic hits bottom; the "trans-contextual individual" agnoizes over his double bind until, in a supportive environment, he finally makes it to creativity. But since Bateson himself argues that "no amount of rigorous discourse of a given logical type can 'explain' phenomena of a higher type,"[39] it is very likely that the deliberate triggering of Learning III can take place only by way of traditional archaic practices. In other words, the intellect generates yearnings for a larger type of mental experience, a wider consciousness, but it can only take you to the edge of such an experience. The actual perception of subject/object merger, of the world as being totally alive and sensuous—in short, the "God-realization"—is a purely visceral event. If Bateson is not advocating traditional practices, it is unclear how anyone can have this insight; and if he *is* advocating them, then Learning III is going to be fraught with the same sorts of political problems that these practices typically bring in their wake.

What are these problems? The major one is that of transference, blind devotion to the guru or teacher, which seems almost inevitably to accompany the experience of "having one's mind blown." In all such practices, the techniques of meditation, breathing, chanting, and so forth serve to reduce external sensory input so that ego-consciousness starts to take itself as its own object of scrutiny. To use cybernetic terminology, the program (Learning II) goes into overload; it begins to appear to itself as an arbitrary construct. The individual loses his or her sense of reality, which now takes on a kind of floating quality. Terror may set in, for the ego perceives itself as dying and cannot imagine what will survive its dissolution. It is at this point that the guru, or teacher, becomes crucial, because his existence is living proof that something does in fact survive. His goal is to help the novice negotiate the Abyss, the gap between mind and Mind. Finally, the wall between conscious and unconscious breaks down completely, and the sensation is that of being swamped, of being carried along in an ocean of God-realization. This perception is experienced as one of immense clarity, of suddenly waking up to what one feels is fully real. If the process is successful, the student who makes it to Learning III continues to experience a gap between mind and Mind, but now without terror or ecstasy. Instead, he sees ego-consciousness as a tool: useful, but hardly anything to stake one's life on. He knows that reality is much larger than this; that, as Laing put it, the ego can and should be the servant of the divine rather than its betrayer.

What next? What do you do with God once you've found Him? As the phrase "awakening to ecstasy" suggests, the student's life is irrevocably altered. The sensation is that of emerging from darkness for the first time, and knowing now (as in the Platonic parable of the cave) how truly unaware one's previous "awareness" really was. All of one's feelings can easily become focused on the teacher, now seen as a father writ large, the person who made this liberation possible. We have all met the person who is constantly quot-

287

ing his or her therapist ("Well, *Tania* says that. . . ."), a tendency that is a variety of guruism. Direct guruism is much worse; it is adulation of the blindest sort, the very opposite of freedom. What began as liberation ends in worship; the believer's life is no longer his or her own. The guru's word is law.

And what *is* the guru's word? What is he actually teaching? Usually, that his word *is* law! It would be bad enough if the process ended with adulation of the teacher, and that was that. The real problem is that the guru, especially in the context of a manipulative society, has a hidden agenda, and it is more often power than money. So the student gets deprogrammed, has his or her Learning II stripped away, sees ultimate reality, and before the dust settles, as Michael Rossman puts it, "is given a full prefab[ricated] structure to put in its place." But there is, Rossman adds, a big difference between worshiping the mystery that is revealed, and worshiping the revealer and his framework. There is always a metacurriculum with a guru, and it is totalitarian—hardly the type of *solve et coagula* that the alchemists had in mind.[40]

Nor is guruism the type of personality redefinition that Bateson had in mind, and it seems to me that an important potential safety valve is suggested in his work. The concluding pages of *Mind and Nature* reveal that just before his death, Bateson was starting to move toward a theory of aesthetics which could have provided a framework of sacredness or beauty for the evolution from ego-consciousness to something larger. Conceivably, such a theory could have been an open door to the planetary culture described above; it now remains for others to develop. Yet even if an appropriate theory of aesthetics is developed, it is not clear how it could have a serious political impact. It would have to be, as Bateson's own work is, an experience, a mode of living, not a formula. This involves personal choice, in other words; a *politics* of self-realization may not be possible. A theory of aesthetics might be valuable to the individual explorer who is making the journey from contemporary science to holism; ideally, it would enable him or her to make that journey without falling prey to guruism. But one of the strengths of Bateson's work is its relational quality; it is not enough to discover the "vast ecology" for yourself alone. The converted alcoholic includes in this ecology the other members of Alcoholics Anonymous and their common struggle. This social emphasis is very positive in the case of AA; the problem arises when the organization is not so benevolent, not interested in health or freedom but in political aggrandizement (usually in the name of health and freedom). Unfortunately, the desire to exert power over others is the rule rather than the exception, and it is hard to see how any theory of aesthetics would be able to influence or control the phenomenon of guruism writ large. We need a safety valve that allows the process of Learning III to occur but not get out of hand; and since no one has managed to come up with anything like this, I feel the need to make a few additional comments on the dangers of Learning III and its possible political implications. Strictly speaking, the following discussion is neither a critique of Bateson personally nor of his work. Neither he nor it, as I suggested earlier, had or have any sympathy whatever for the right-wing cultism that Learning III is currently generating. Rather, it reflects my own fears that no holistic philosophy to date has managed to provide adequate safety valves with respect to the Learning III process, and thus that any discussion of the process has to be accompanied by a warning note.

If the danger of Learning III is one of transference, we should not be surprised at the mental colonization being practiced by numerous right-wing cults, especially in the United States.[41] In his book on television, former advertising executive Jerry Mander has done a fine job of explicating the process in the case of Werner Erhard's organization *est*, though he is quick to point out that his selection of *est* as an example is virtually arbitrary.[42] *Est's* approach includes many of the classic techniques of Zen or yogic training—meditation, visualization, the deliberate reduction of sensory stimuli—and the result is not liberation, but a forest of clones. *Est* followers tend to dress alike, talk alike, and use a jargon eerily reminiscent of Batesonian holism ("Mind," "context," "programming," and so on). The talk is all of "taking responsibility for oneself," but the disciples have an ambiance that bears an uncanny resemblance to that of Erhard, and have been dubbed "talking parking meters" by the California press. The phenomenon of *est*, writes Rossman, has given us "the spectacle . . . of relatively intelligent people handing over their minds en masse";[43] and this willing abandonment of critical faculties on the part of his followers has enabled Erhard to expand his base of operations significantly. His enterprise now includes such public relations gimmicks as a fraudulent "hunger project," and Erhard's appointment as a professor of "context" (!) at Antioch's Holistic Life University. What *est* teaches, from a political viewpoint, is pure rubbish (victims always choose their fate, presumably even babies napalmed in Vietnam), and need not concern us here. The real cause for concern is that despite their widespread popularity, Erhard, the Reverend Moon (Unification Church), L. Ron Hubbard (Church of Scientology), and their ilk are relative amateurs. Most people have steered clear of these organizations, and the political structure of industrial society has up to this point been untouched by these Learning III racketeers. But we have not seen the last of such false messiahs, and sooner or later one of them, with government encouragement, might catch on as a mass phenomenon. Erhard has tried to court people in positions of influence and power, but without any known success. In Nazi Germany, those adept at manipulating the unconscious did not need to court the government; they *were* the government. "Hitler," wrote the German sociologist Max Horkheimer shortly after the war, "appealed to the unconscious in his audience by hinting that he could forge a power in whose name the ban on repressed nature would be lifted."[44] Current conditions hardly rule out the possibility of a repeat performance.

The specter of fascism, of course, is often invoked by those who want to rationalize their opposition to political change, but I sense that in this case it is no idle threat. We are talking about reviving the psychic underbelly not within the context of a traditional society that is still in touch with its grounding, but within the framework of a mobile, rootless, high-technology, sexually repressed, mass society. The parallel with Germany after World War I is close, for that was a society in which myth and symbol, sexuality and occultism, the "natural" and the nonrational, were deliberately cultivated as antidotes to an artificial, over-intellectualized, bureaucratized way of life.[45] The psychic energy thus made available was enormous, and was brilliantly colonized by the Nazis at immense rallies held in Nuremburg and Munich—mimetic performances complete with giant swastikas and klieg lights—for their own political purposes. "The people" were hardly the winners in this officially sanctioned "liberation" from their own repression.

It was the danger of such mysticism which Immanuel Kant had in mind when he called reason (ego-consciousness) "the highest good on earth," "the ultimate touchstone of truth"; and commenting on this statement in 1945, Lucien Goldmann wrote:

The last twenty-five years have shown us how penetrating Kant's vision was and how close are the ties which link irrationalism and the mystique of intuition and feeling with the suppression of individual liberties.[46]

Given enough social and economic chaos, and the increasing number of self-proclaimed gurus, there is every reason to keep in touch with our old deutero-learning.

The link between the nonrational and state power in general depends upon an elitism that is implicit in most guruism. Most, but not all. The shaman of traditional cultures spoke the voice of God (when in trance) and that was that. He generally made no bid for secular control. But in a civilization that has lost its own roots, teachers of Learning III do not merely charge high fees for their services; some also, like Erhard, want power of the most absolute sort. Their claim to it lies precisely in the distinction between "wakers" and "sleepers." There is a spiritual pecking order here, a separation of orthodox from heterodox, the self-realized from those who have not yet "awakened to ecstasy" and who may never do so. William Irwin Thompson recently argued that ego-consciousness being what it is, "we should trust no policy decisions which emanate from persons who do not yet have [the] habit [of Mind]. We must not let anyone near the political process who has not stepped out of small mind and encountered the fullness of Being."[47] Thompson's is an important point, but what is the alternative? Who is the "we" Thompson is referring to? As he himself states in the next breath:

> The difficulty with this idea is that it is a theory of elites. The Elite become the new policy-makers, the new politicians, the new human-ity, the new homo sapiens. . . . This globalist elite could then make a rapprochement with the multinational corporation executives to introduce a new authoritarian world-order.[48]

Holism, in short, could become the agent of tyranny, but in the name of Mind, Learning III, or (God help us) God. It was not for nothing that Orwell once remarked that when fascism finally comes to the West, it will do so in the name of freedom.

Reflecting on the mechanical philosophy of the Scientific Revolu-tion, Alfred North Whitehead once remarked that with its formula-tion the West found itself in the grip of an idea it could live neither with nor without. Surely, the same thing can be said of Learning III, or *mimesis* in general. The disembodied consciousness of the modern era is barbaric; it is integral to the landscape described in the Introduction. But the attempts to escape such a world by institutionalizing Learning III have often been no less barbaric. The key phrase here is, "such a world". Even total *mimesis* is not barbaric in a bicameral, or totally primary-process world, as Julian Jaynes has demonstrated.[49] The problem arises when worlds collide. As Reich realized, industrial democracy is dry tinder for fascism and the irrational precisely because it *is* so sterile, so Eros-denying, and because it has been with us now for centuries. Neither in society nor in a single individual can one suddenly remove such fantastic blockage and expect the reaction to be one of smooth and sensible readjustment. We thus confront a choice that must be made and yet cannot be made: the awakening of an entire civilization to its repressed archaic knowledge. It is not likely that the mental world view of Cartesian deutero-learning, which includes traditions such as social democracy, secular humanism, and enlightened (or vulgar) Marxism, can make this choice in an intelligent way, because these traditions insist that Mind, or Being, is an obscurantist concept. But as one atypical observer, Ernst Bloch, pointed out in 1931, the Left in Germany was ignoring developments occurring in primitive and utopian trends, thereby leaving a whole territory free for the Nazis to occupy.[50] Repression works only up to a point; utopian longings stir even in the most subjugated individual, and fascism recognizes those longings and manipulates them to its advantage. As indicated above, the celebration of nature versus artifice is a central tenet of fascist ideology. The revolt of "natural man" versus technology, the destruction of spontaneity, and the domination of nature are all foolishly ignored by mainstream or "progressive" politics; but when these issues do become central, politics can take on frightening dimensions. "In this light," writes Max Horkheimer, "we might describe fascism as a satanic synthesis of reason and nature—the very opposite of that reconciliation of the two poles that philosophy has always dreamed of."[51]

Nevertheless, I believe our evolution toward Learning III is inevitable, and if so, the real question becomes: What is a safe context for it? What institutional structures would be beneficial to its healthy flowering? To some extent, this question has already been answered in our earlier discussion of planetary culture. A decentralized set of autonomous regions is the very opposite of the rootless, mass society that makes Learning III such volatile stuff. Self-determination, strong local community ties, neighborhood spirit—all of these would break down the globalist monolith and thus serve to contain any revival of the archaic which threatened to turn into a mass movement. The whole process of balkanization has its problems, of course, but I doubt that global totalitarian *mimesis* is one of them. The Third Reich, for example, was hostile to regional sentiment. It was a nation-state ultimately made possible by Bismarck's forced unification of the small German states, and it countered regional sentiments with its policy of *Lebensraum,* which aimed to force neighboring territories into a centralized, German world-order. By itself, decentralization cannot eliminate guruism, but it can certainly limit its influence. A rooted society is protection not only against alienation—which is the product of the attempt to control everything—but also against its opposite, which involves the complete loss of control.

What shall such a society be rooted in? Traditionally, regional or community politics was the politics of ethnicity. One had loyalty to one's clan, kinship system, race, or language group. It is doubtful that the ethnic model can work anymore in a world that has seen several centuries of global communication and fairly violent culture contact. And this may be all to the good, for regional ethnicity can easily turn into a provincial type of ethnic chauvinism, which finally results in a *narrowing* of human possibilities. Cosmopolitanism is still a fine ideal, and thus the need is not merely for rootedness, but for a rootedness that also encourages planetary interdependence and cultural exchange. Given the disruption of familial and local bonds over the last few centuries, many people in Western industrial societies now seek new sources of communitarianism which do not also threaten to close off their mental horizons. There are no easy answers forthcoming, and there may be no way out of this dilemma. *In situ* cultures are not congenial with the "Gutenberg galaxy."

The irreconcilability of planetary versus globalist world views, or what has been termed ecosystem versus biospheric cultures, has of late been elaborated upon by ecologist Raymond Dasmann.[52] The former depend on the local ecosystem for food and materials, and environmental protection is guaranteed through religious belief and social custom. Such people, the American Indians for example, have (or had) awesome in-place skills. They know the local animal species, the meaning of the slightest shift in the wind, and have a rich lore of herbs and their preparation. Their lives are tailored to an optimum relationship with their particular region, or what Peter Berg terms *bio*region, in which "culture is integrated with nature at the level of the *particular ecosystem* and employs for its cognition a body of metaphor drawn from and structured in relation to that ecosystem."[53] Recent research indicates that historically, such people lived relatively abundant lives, and did so with far less work than we do today.[54] Biospheric people, on the other hand, take the entire globe as their province, drawing on vast networks of trade and communication. There is nothing place-specific about their knowledge, and they can do whatever they want to any particular region they choose. Whereas ecosystem people might deal with water shortages by building rooftop collectors and storage tanks, attending to the local vegetation, and maybe holding a rain dance or two (all at trivial economic and ecological expense), biospheric people build giant dams and canal systems that disrupt the environment and cost millions. As we know, in order for biospheric people to have what they want, ecosystem people must sell out or, as is more typically the case, be wiped out. But the truth, says Dasmann, is that the biospheric people are ultimately the losers in this global shell game, because their "victory" involves the loss of a vast network of skills and habits that have enabled man to sustain himself on the planet for millennia. The economies of biospheric societies are not sustainable and are now in chaos, argues Dasmann; American resource policies are an example for the rest of the world of what *not* to do. "I would propose," he concludes, "that the future belongs to those who can regain, at a higher level, the old sense of balance and belonging between man and nature." Rootedness, in short, must become biotic, not merely ethnic, and Dasmann has contructed a map of the "biotic provinces of the world," showing what political boundaries would be like if they followed the lines of natural geography and species density variation.[55] The bioregional model of Berg and Dasmann is posited on the distinction between occupying a region and inhabiting it; or, for us now, *re*inhabiting it. "*Reinhabitation*," write Berg and Dasmann,

> means learning to live-in-place in an area that has been disrupted and injured through past exploitation. It involves becoming native to a place through becoming aware of the particular ecological relationships that operate within and around it. It means understanding activ-

ities and evolving social behavior that will enrich the life of that place, restore its life-supporting systems, and establish an ecologically and socially sustainable pattern of existence within it. Simply stated it involves becoming fully alive in and with a place. It involves applying for membership in a biotic community and ceasing to be its exploiter.[56]

It is a fine vision, and the authors may be right when they argue that "living-in place... may be the only way in which a truly civilized existence can be maintained."[57] But whether the rootless, urbanized people of Europe and North America can now create a source of identity around biotic provinces and bioregional loyalties that were largely obliterated centuries ago is an open question.

And yet, what other choice do we have? Learning III will continue to gain momentum, and the most crucial political issue of the twenty-first century may be how to provide it with a proper context. As we noted earlier, Learning III has been tapped in all traditional cultures by certain techniques of initiation. That this process did not get out of hand was not only a function of having a small-scale, decentralized way of life. We saw that in organizations such as *est*, once a floating reality is obtained the initiators, or gurus, implant their own reality in the person, usually the worship of the guru and his organization. Now it is clearly the case that all tribal, *in situ* cultures have their shamans, and the initiation process led by the shaman is also designed to break down Learning II. But in a world that is rooted in bioregional realities, such as these cultures are, the process does not lead to transference and blind obedience to authority. What develops in the Learning III process is not adoration of the shaman, but of the mystery he makes manifest: the God within, and the ecosystem that reflects it. This was the final lesson Carlos Castaneda learned in his initiation at the hands of don Juan, and it is the message of all nature-based religions.[58] It generates what social critics Jerry Gorsline and Linn House describe as "a science of the *concrete*, where nature is the model for culture because the mind has been nourished and weaned on nature."[59] In short, it is my guess that preservation of this planet may be the best guideline for *all* our politics, the best context for *all* our encounters with Mind or Being. The health of the planet, if it can be successfully defended against the continuing momentum of industrial socialism and capitalism, may thus be the ultimate safety valve in the emergence of a new consciousness. And it is only in such a world, I believe, that the Cartesian paradigm can be safely discarded, and human beings begin living the lives they were meant to live all along: their own.

Regardless of its duration as a political entity, every civilization, like every person, is a message—makes a single statement to the rest of the world. Western industrial society will probably be remembered for the power, and the failure, of the Cartesian paradigm.

When I was a boy, the Cartesian paradigm seemed infallible to most Westerners, successful without parallel in the history of the human intellect. This way of life was celebrated in space programs, rapid technological innovation of all sorts, and books with titles such as *The Endless Frontier* and *The Edge of Objectivity*. By the mid-1960s, it was becoming clear to many that science was, in fact, an ideology; and from that point it was a short step to the recognition that it was not a very healthy ideology at that.

It is very likely that the next few decades will involve a period of increasing shift toward holism, Batesonian or otherwise. As scientific civilization enters its period of decline in earnest, more and more people will search for a new paradigm, and will undoubtedly find it in various versions of holistic thinking. If we are lucky, by 2200 A.D. the old paradigm may well be a curiosity, a relic of a civilization that seems millennia away. Jung, Reich, and Bateson especially, have each helped to point the way to a reenchanted world in which we can believe. Once again, the secular would be the handmaiden of the sacred, but with at least some ego-consciousness left intact. Yet from the vantage point of an extended time scale, one wonders if an ancillary arrangement will be enough. The period from Homer to the present is not even 3,000 years—a mere blink of the eye in anthropological terms; the last four hundred years may prove to be only the most aggravated phase of a single evolutionary episode. If so, the next phase in our evolution, that of self-conscious *mimesis*, may actually be a transitional one. Reenchanting the world, even nonanimistically, may ultimately necessitate the end of ego-consciousness altogether. The French psychiatrist Jacques Lacan has argued that the ego is a paranoid construct, founded on the logic of opposition and identity of self and other. He adds that all such logic, which is peculiar to the

West, requires boundaries, whereas the truth is that perception, being analogue in nature, has no intrinsic boundaries.[60] As our epistemology becomes less digital and more analogue, boundaries will begin to lose definition. Ego, character armor, "secondary process" will start to melt. We may then begin to move back to what Robert Bly calls "Great Mother culture," to cosmic anonymity, a totally mimetic world.[61]

If such is indeed our fate, it is nevertheless the case that the transformation will not happen overnight. As I have suggested above, a too rapid devolution would probably spell untold disaster. If we are lucky, the interim period will involve a revival of the unconscious, and the development of relational or holistic perception, but with enough awareness of the subject/object distinction so as to prevent untoward events. We shall need to keep our wits about us, in short, and that means the retention of some ego-consciousness. But ultimately, ego-consciousness may not be viable for our continuation on this planet. The end of alienation may lie not in the reform of the ego, or in complementing it with primary process, but in its abolition.

There is a famous papyrus in the Berlin Museum, No. 3024, titled *Rebel in the Soul,* and dating somewhere from 2500 to 1991 B.C. This was the so-called Intermediate Period of Egyptian history, between the Old and Middle Kingdoms, a time of total social breakdown, widespread chaos, and disruption. It reflects an age similar to our own, in which old values had collapsed and new ones not yet taken their place. The document records something unheard of in bicameral culture—an identity crisis. Its author is preoccupied with the meaning of life, his self (ego), the conflict between reason and emotion, and possible suicide. The papyrus is hardly typical of hieroglyphic texts, and many near Eastern experts regard it as the only ancient Egyptian document of its kind. Its emergence during the Intermediate Period is evidence for Julian Jaynes's central argument, that when the subject/object distinction did occur in ancient times, its function was a crisis function, the sounding of an extreme alarm. What I have tried to argue in the present work is that since 1600 A.D., and most visibly since the Industrial Revolution, the West has been in a perpetual crisis, an unstable society in a state of extreme alarm. Thus modern schismatic consciousness is regarded as normal, but the times have not been "normal" for centuries. The correspondence with the Egyptian Intermediate

Period is clear here, but with a peculiar twist. The lonely author of *Rebel in the Soul* was probably an enigma to his contemporaries, in that he *found* his ego, whereas we tend to regard psychotics today as enigmatic for having *lost* it. In other words, we may now be moving toward health, whereas the Egyptians of the Intermediate Period were at least temporarily moving toward pathology. Reading the text, we cannot help but recognize a modern voice; to our ears, for example, his words are often heroic. "Brother," says his soul to him, "as long as you burn you belong to life." This is effectively what Teiresias tells Odysseus when the latter visits him in Hades and asks the prophet to show him the way home and put an end to his restless search. But Teiresias is *disapproving* of this twenty-year search for the Self; he hints to Odysseus that a life that is equivalent to "burning" might be well worth giving up.[62] Contemporary existentialist philosophers such as Rollo May, by contrast, have made a career out of the notion that such anxiety, and preoccupation with identity, is a sign of health. They never seem to grasp that we, like the author of *Rebel*, live in times so crazy that *Angst* and vitality get mistaken for one another. Surely, as Christopher Hill would say, ours is a world turned upside down.[63]

The end of ego-consciousness hardly necessitates the end of life, culture, or meaningful human activity. The existentialist position of equating meaning with anxiety can only be maintained by ignoring the major part of man's history on this planet. Ego-consciousness, let alone the tradition of modern individualism, is a phenomenon with a comparatively short history; it is hardly essential for human survival or for a rich human culture, and may ultimately be inimical to both. Thus ecologist Paul Shepard has pointed out that it was a devolution in the Neanderthal brain which gave rise to the smaller-brained Cro-Magnon man (ca. 40,000 B.C.) and Aurignacian civilization (ca. 23,000 B.C.), a period remarkable for cave painting, the invention of nearly two hundred kinds of tools, and a general burst of cultural activity.[64] As Julian Jaynes has pointed out, the neurology of consciousness is hardly set for all time. We may be on the verge of such a period of dynamic devolution, in which what is emerging is not merely a new society, but a new species, a new type of human being. In the last analysis, the present species may prove to be a race of dinosaurs, and ego-consciousness something of an evolutionary dead end.

"When you bring your flesh to rest," the author of *Rebel* is told by his soul,

298

And thus reach the Beyond,
In that stillness shall I alight upon you;
then united we shall form the Abode.

Who shall live in that Abode, and how they shall live, will be for
future historians to say. But given such a world, they may not feel
the need to do so.

Notes

Introduction: The Modern Landscape

1. Morris Berman, *Social Change and Scientific Organization* (London and Ithaca, N.Y.: Heinemann Educational Books and Cornell University Press, 1978).

2. Russell Jacoby, *Social Amnesia* (Boston: Beacon Press, 1975), p. 63.

3. Herbert Marcuse, *One-Dimensional Man* (Boston: Beacon Press, 1964), pp. 9, 154.

4. Studs Terkel, *Working* (New York: Avon Books, 1972).

5. Richard Sennett and Jonathan Cobb, *The Hidden Injuries of Class* (New York: Vintage Books, 1973), pp. 168ff.

6. The elaboration of this process is perhaps the greatest contribution of the Frankfurt School for Social Research, whose most familiar representative in the United States was Herbert Marcuse. A summary of their work may be found in Martin Jay, *The Dialectical Imagination* (Boston: Little, Brown, 1973). On the popular level, Vance Packard has provided much evidence for this view of the totally manipulated life in books such as *The Status Seekers*, *The Hidden Persuaders*, and several others.

7. Joseph A. Camilleri, *Civilization in Crisis* (Cambridge: Cambridge University Press, 1976), pp. 31–32. The incredible emphasis on sexual technique, as opposed to emotional content, is reflected in the voluminous proliferation of sex manuals in the last fifteen years, by now a multimillion dollar business.

8. R. D. Laing, *The Divided Self* (Harmondsworth: Penguin Books, 1965; first publ. 1959).

9. For a report on a study of performance anxiety among first graders, which was carried out by William Kessen of Yale University, see Barbara Radloff, "The Tot in the Gray Flannel Suit," *New York Times*, 4 May 1975. "You have to play by the rules of the game if you are going to survive," she states, "in a corporation, or in first grade." The distinction between inner vitality and outer sterility which is familiar to all high school students formed a persistent theme in the rock music of the 1950s. Chuck Berry's

songs, such as "School Days" and "Sweet Little Sixteen," are perhaps the prototype.

10. Camilleri, *Civilization in Crisis*, p. 42. Information such as this can be collected, at this point, by merely reading daily newspapers and popular journals. My own sources include: *Newsweek*, 8 January 1973 and 12 November 1979; *National Observer*, 6 March 1976; *San Francisco Examiner*, 24 March 1977 and 10 July 1980; *San Francisco Chronicle*, 29 March 1976 and 10 September 1979; *New York Times*, 16 March 1976; *Cosmopolitan*, September 1974; and a general survey of such articles provided in John and Paula Zerzan, "Breakdown," which was published in abridged form in the January 1976 issue of *Fifth Estate*. The quotation from Darold Treffert is from this pamphlet. For an extended critique of American drug use, see Richard Hughes and Robert Brewin, *The Tranquilizing of America* (New York: Harcourt Brace Jovanovich, 1979).

11. According to a 1972 Finnish study, Poland, the Soviet Union, and Hungary are respectively first, second, and third in world per capita consumption of hard liquor. See *San Francisco Chronicle*, 8 September 1978. My information on French and German suicides comes from a 1979 report of San Francisco's Pacific News Service by Eve Pell, "Teenage Suicides Sweep Advanced Nations of the West."

12. Dr. Edward F. Foulks, a medical anthropologist at the University of Pennsylvania, has argued that madness may be a way by which the human species protects itself in such times of crisis, and hence that psychosis may be a form of cultural avant-garde (see the report on his work in the *New York Times*, 9 December 1975, p. 22, and the *National Observer*, 6 March 1976, p. 1). Much of the work of R. D. Laing points in this direction, and it has been a theme in a number of Doris Lessing's novels. See also Andrew Weil, *The Natural Mind* (Boston: Houghton Mifflin, 1972).

13. Robert Heilbroner, *Business Civilization in Decline* (New York: Norton, 1976), pp. 120–24.

14. Willis W. Harman, *An Incomplete Guide to the Future* (San Francisco: San Francisco Book Company, 1976), chap. 2.

CHAPTER 1. *The Birth of Modern Scientific Consciousness*

1. Christopher Marlowe, *The Tragedy of Doctor Faustus*, ed. Louis B. Wright and Virginia A. LaMar (New York: Washington Square Press, 1959), p. 3; reprinted by permission of Simon and Schuster.

2. Francis Bacon, *New Organon*, Book I, Aphorism XXXI, in Hugh G. Dick, ed., *Selected Writings of Francis Bacon* (New York: The Modern Library, 1955). This and subsequent excerpts printed with the permission of Random House, Inc.

3. "Pure" historians of ideas have tended to see Bacon as irrelevant, or

even detrimental to the growth of modern science, partly due to their own reaction against Marxist historians such as Benjamin Farrington (*Francis Bacon: Philosopher of Industrial Science* [New York: Collier Books, 1961; first publ. 1949]), who see Bacon as a cultural hero. The most extreme expression of this is C. C. Gillispie, *The Edge of Objectivity* (Princeton: Princeton University Press, 1960), pp. 74–82.

4. In addition to Farrington's work, good discussions of this point can be found in two books by Paolo Rossi: *Francis Bacon*, trans. Sacha Rabinovitch (London: Routledge & Kegan Paul, 1968), and *Philosophy, Technology and the Arts in the Early Modern Era*, trans. Salvator Attanasio (New York: Harper Torchbooks, 1970). See also Christopher Hill, *Intellectual Origins of the English Revolution* (London: Panther Books, 1972), chap. 3.

5. Bacon, *New Organon*, Book I, Aphorism LXXIV.

6. Ibid., Aphorism XCVIII.

7. There is, of course, a large literature comparing Eastern and Western science and modes of thought. A fine one-volume summary is Joseph Needham, *The Grand Titration* (London: Allen & Unwin, 1969).

8. This and all of the quotations from Descartes are taken from his *Discourse on Method*, trans. Laurence J. Lafleur (Indianapolis: The Liberal Arts Press, 1950; original French edition, 1637).

9. A spirited discussion of this disparity can be found in Pierre Duhem, *The Aim and Structure of Physical Theory*, trans. Philip P. Wiener (New York: Atheneum, 1962; original French edition 1914), chap. 4.

10. Descartes, *Discourse*, p. 12.

11. A. R. Hall, *The Scientific Revolution* (Boston: Beacon Press, 1956), p. 149. My earlier statement, that for Descartes "all nonmaterial phenomena ultimately have a material basis," is thus not strictly true. For Descartes, *res cogitans* and *res extensa* were distinct entities; it was Descartes' disciples who made mind epiphenomenal and attempted to swallow up the former by the latter—as is commonly done in science today. Despite Descartes' original sophistication, mainstream Cartesianism came to be identified with materialist reductionism.

12. I am adopting the distinction between critical and dialectical reason made by Norman O. Brown in *Life Against Death* (Middletown, Conn.: Wesleyan Univ. Press, 1970; orig. publ. 1959).

13. The best one-volume discussion of Galileo's work, to my mind, is Ludovico Geymonat, *Galileo Galilei*, trans. Stillman Drake (New York: McGraw-Hill, 1965).

14. Piaget has reported his findings in a large number of works. The latest study is *The Grasp of Consciousness*, trans. Susan Wedgwood (Cambridge: Harvard University Press, 1976). To preclude any confusion in the following discussion and in Chapter 2, I should state that I am not an Aristotelian and am not suggesting a return to the Thomistic synthesis of

the Middle Ages. Rather, my interest in Aristotle here and in Chapters 2 and 3 is related to the presence of participating consciousness in his work. There is obviously more to Aristotle than this, including his laws of logic and noncontradiction which run directly counter to the notion of participation, and which constitute the basis of much contemporary scientific reasoning to this day.

15. It should be clear that to enter the world of modern science is to enter a world of abstractions that violate everyday observations. From 1550 to 1700 Europe did enter wonderland, as surely as Alice did when she fell down the rabbit hole. But the fall, I would maintain, was not clean. Certainly the dominant culture of science and technology linked to the creation of material wealth is the other end of the drop, and students training for positions in that culture are quickly reeducated to the Newtonian/Cartesian/Galilean mode of perception; but privately, and emotionally, we still operate in the common-sense world of immediate experience—a world in which objects naturally fall to the center of the earth, and all motion obviously requires a mover. We even retain traces of animism, over the years developing an almost personal relationship with a favorite chair or lamp, even though we "know" it is nothing but wood or metal.

16. Oskar Kokoschka, *My Life,* trans. David Britt (London: Thames and Hudson, 1974), p. 198.

17. Bertolt Brecht, *Galileo,* trans. Charles Laughton and ed. Eric Bentley (New York: Grove Press, 1966, from the English edition of 1952), p. 63. Reprinted by permission of Indiana University Press.

18. Actually, one wonders. Rochefoucauld related an incident that occurred in the latter half of the eighteenth century, in which a Norfolk clergyman, being examined for his doctorate at Cambridge, was asked whether the sun went around the earth or the earth around the sun. "Not knowing what to say, and wanting to make some reply, he assumed an emphatic air and boldly exclaimed: 'Sometimes the one, sometimes the other.'" Amazingly enough, he was awarded the degree. See G. E. Mingay, *English Landed Society in the Eighteenth Century* (London: Routledge & Kegan Paul, 1963), p. 137.

19. The date of publication of the *Principia* is commonly given as 1687, but H. S. Thayer, in *Newton's Philosophy of Nature* (New York: Hafner, 1953), p. 9n, cites 1686 as the correct year of publication of the first edition.

20. Quoted in Thayer, p. 54; reprinted by permission of the Macmillan Publishing Co., Inc.

21. Ibid., p. 45; reprinted by permission of the Macmillan Publishing Co., Inc.

22. Positivism probably received its earliest formulation in the work of Marin Mersenne (see below, Chapter 3). A fully modern statement of it is contained in Roger Cotes's Preface to the second edition of the *Principia,* reprinted in Thayer, pp. 116–34, esp. p. 126.

23. Alfred North Whitehead, *Science and the Modern World* (New York: Mentor Books, 1948; orig. publ. 1925), p. 55.

24. N. O. Brown, *Love's Body* (New York: Vintage Books, 1966), p. 139.

25. Peter Berger, "Towards a Sociological Understanding of Psychoanalysis," *Social Research* 32 (Spring 1965), 32. The classic statement of the sociology of knowledge is Karl Mannheim, *Ideology and Utopia*, trans. Louis Wirth and Edward Shils (New York: Harvest Books, reprint of 1936 edition).

CHAPTER 2. *Consciousness and Society in Early Modern Europe*

1. Ernest Gellner, *Thought and Change* (Chicago: University of Chicago Press, 1964), p. 72.

2. Cf. Carlo M. Cipolla, *Before the Industrial Revolution* (New York: Norton, 1976), pp. 117–18.

3. I shall discuss the problem of the sociology of knowledge and radical relativism briefly at the conclusion of this chapter, and in detail in Chapter 5. As for the issue of causality, the reader should be aware that much of the literature in the history of science revolves around a debate over the role of "external" factors in the rise of modern science versus the role of "internal" factors (i.e., factors arising from social influence as opposed to those that are rooted in the material of scientific development itself). Not surprisingly, the debate has never been resolved, for it depends entirely on the artificial mind-body dichotomy of the modern era. As discussed in Chapter 3, this split was not experienced by pre-modern society. Once the dichotomy is recognized for what it is, the "externalist-internalist" argument evaporates.

For some of the more classic essays on the subject, consult the following anthologies: Hugh F. Kearney, ed., *Origins of the Scientific Revolution* (London: Longmans, Green, 1964); George Basalla, ed., *The Rise of Modern Science* (Lexington, Mass.: D. C. Heath, 1968); Leonard M. Marsak, ed., *The Rise of Science in Relation to Society* (New York: Macmillan, 1964).

4. E. A. Burtt, *The Metaphysical Foundations of Modern Science*, 2d ed. (Garden City, N.Y.: Doubleday, 1932).

5. John Donne, "An Anatomie of the World: The First Anniversary," in *Donne*, ed. Richard Wilbur (New York: Dell, 1962), pp. 112–13 (reprinted with the permission of Oxford University Press). Pascal is quoted in the original ("Les silences des espaces éternels m'effrayent") in W. P. D. Wightman, *Science in a Renaissance Society* (London: Hutchinson University Library, 1972), p. 174.

6. As might be expected, the literature on feudalism, the Commercial Revolution, and the transition to capitalism is so vast as to defy any attempt at bibliography. For descriptions of these processes I have used the following works: Fernand Braudel, *The Mediterranean and the Mediterranean*

World in the Age of Philip II, vol. 2, trans. Siân Reynolds (New York: Harper & Row, 1972); Pierre Jeannin, *Merchants of the 16th Century*, trans. Paul Fittingoff (New York: Harper & Row, 1972); Carlo M. Cipolla, *Before the Industrial Revolution;* Immanuel Wallerstein, *The Modern World-System* (New York: Academic Press, 1975); and J. U. Nef, *Industry and Government in France and England, 1540–1640* (Philadelphia: American Philosophical Society, 1940).

7. Alfred von Martin, *Sociology of the Renaissance* (New York: Harper Torchbooks, 1963; orig. German edition 1932), pp. 14, 21; these and subsequent excerpts reprinted with permission of the publisher. The transition from sacred to secular number (cabala to bookkeeping, for example) was part of this general process, and is discussed briefly in Chapter 3.

8. Ibid., p. 40.

9. Mircea Eliade, *The Myth of the Eternal Return; or, Cosmos and History,* trans. Willard R. Trask (Princeton: Princeton University Press, 1971; orig. French edition 1949), and von Martin, *Sociology of the Renaissance,* p. 16.

The reader should note that linear time was experientially alien, but not officially alien, to the medieval mind. Official Christian time of the Middle Ages was linear, in that it was believed that there was a particular point at which the world had been created and that it was now moving toward the Second Coming (which was, however, a re-creation). Similarly, each individual was moving from his own birth to his death and (ideally) salvation. To the extent that Christian culture adopted the framework of Jewish eschatology, then, it did think in terms of linear time. However, Eliade and von Martin are not referring to biblical or official conceptions of time, but to time as it was experienced in the daily fare of life. What was *felt* was indeed cyclical: the sun rises and sets, seasons follow each other year in and out, planting is followed by harvest, and even church holidays can be counted on to recur faithfully each year. There are probably several strands of thinking about time in the Middle Ages, but I believe Eliade and von Martin have captured the dominant mode of consciousness.

10. Lynn White, Jr., *Medieval Technology and Social Change* (London: Oxford University Press, 1964), p. 125.

11. For literature on the scholar and the craftsman see note 3 to this chapter. Especially relevant are the articles by A. R. Hall and E. Zilsel in Kearney, *Origins of the Scientific Revolution,* pp. 67–99, and Paolo Rossi, *Philosophy, Technology and the Arts in the Early Modern Era,* trans. Salvator Attanasio (New York: Harper Torchbooks, 1970).

The discussion below generally applies to the middle-class artisan, or master craftsman, rather than the lowest level of artisan. The former, like the military engineer, had some nonvocational education, while the latter usually did not. By 1600, there were already class divisions between apprentices, journeymen, and master craftsmen.

12. A phrase that was becoming popular in the late sixteenth century.

William Gilbert paraphrased it in the Preface to his book *De Magnete* (On the Magnet) of 1600.

13. Rossi, *Philosophy, Technology and the Arts*, pp. 30–31.

14. Ibid., p. 42.

15. Ibid., p. 112. The sketch of Galileo, Tartaglia, and the scholar-craftsman merger given below is based on the following sources: Ludovico Geymonat, *Galileo Galilei*, trans. Stillman Drake (New York: McGraw-Hill, 1965); Galileo Galilei, *Dialogues Concerning Two New Sciences*, trans. Henry Crew and Alfonso de Salvio (New York: Macmillan, 1914); Gerald Holton and Duane Roller, *Foundations of Modern Physical Science* (Reading, Mass.: Addison-Wesley, 1958); Stillman Drake and James MacLachlan, "Galileo's Discovery of the Parabolic Trajectory," *Scientific American* 232 (March 1975), 102–10; Edgar Zilsel, "The Sociological Roots of Science," in Kearney, *Origins of the Scientific Revolution*, pp. 86–99; Stillman Drake and I. E. Drabkin, trans. and eds., *Mechanics in Sixteenth-Century Italy* (Madison: University of Wisconsin Press, 1969); A. R. Hall, *Ballistics in the Seventeenth Century* (Cambridge: Cambridge University Press, 1952); Stillman Drake, "Galileo and the First Mechanical Computing Device," *Scientific American* 234 (April 1976), 104–13.

16. Galileo Galilei, *Dialogues Concerning Two New Sciences*, p. 1, reprinted with permission of Dover Publications, Inc.

17. This was Imre Lakatos' evaluation of T. S. Kuhn's view of scientific revolutions. See Imre Lakatos and Alan Musgrave, eds., *Criticism and the Growth of Knowledge* (Cambridge: Cambridge University Press, 1970), p. 178.

CHAPTER 3. *The Disenchantment of the World (1)*

1. A number of scholars, including T. S. Kuhn, Claude Lévi-Strauss, Michel Foucault, Roland Barthes, and members of the Frankfurt School (see the Introduction, note 6) have recognized the fallacy of this progress theory of intellectual history, but the epistemological framework(s) that they represent has hardly made a dent in most thinking on the subject. The "asymptotic" view of scientific knowledge is still the common one, and it permeates the media, the universities, and all other institutions of Western culture. This view was perhaps apotheosized by C. P. Snow in his novel, *The Search* (New York: Scribner's, 1958).

2. The study of nonrabbinical Judaism has been the work of Gershom Scholem (*Major Trends in Jewish Mysticism, On the Kabbalah and Its Symbolism*). The gnosticism of Judaism in antiquity has been explored by Erwin Goodenough, *Jewish Symbols in the Greco-Roman Period,* vols. 7–8, *Pagan Symbols in Judaism* (New York: Pantheon Books, 1958), and by Michael E. Stone, "Judaism at the Time of Christ," *Scientific American* 228 (January 1973), 80–87.

Notes

3. Owen Barfield, *Saving the Appearances* (New York: Harcourt, Brace & World, 1965), esp. chap. 16.

4. Julian Jaynes, *The Origin of Consciousness in the Breakdown of the Bicameral Mind* (Boston: Houghton Mifflin, 1976), bk. 1, chap. 3, and bk. 2, chap. 5; and Friedrich Nietzsche, *The Birth of Tragedy*, trans. Francis Golffing (Garden City: Doubleday Anchor, 1956), esp. pp. 84, 107–8. Bennett Simon has an excellent discussion of the Homeric and post-Homeric mentalities in his book *Mind and Madness in Ancient Greece* (Ithaca, N.Y.: Cornell University Press, 1978).

5. The following discussion of Greek consciousness is taken from E. A. Havelock, *Preface to Plato* (Cambridge: Harvard University Press, Belknap Press, 1963), pp. 25–27, 45–47, 150–58, 190, 199–207, 219, 238–39, 261. John H. Finley, Jr., develops the same line of reasoning in his lovely essay, *Four Stages of Greek Thought* (Stanford: Stanford University Press, 1966). Cf. also Simon, *Mind and Madness in Ancient Greece*.

6. Although his analysis of Greek intellectual history can be faulted at several points, Robert Pirsig, with no apparent awareness of Nietzsche's discovery of the reality of participating consciousness, rediscovered it for himself in his autobiographical study of Greek philosophy and, like Nietzsche, went insane as a result (*Zen and the Art of Motorcycle Maintenance* [New York: William Morrow, 1974]). The identical theme is repeated in Doris Lessing's story of one Charles Watkins, a classics professor, in her brilliant novel *Briefing for a Descent Into Hell* (New York: Knopf, 1971), in which Watkins goes mad from his insight and is (like Pirsig) jolted back into nonparticipating consciousness by means of electroshock therapy.

Plato's psychological ideal is perhaps best described in the *Republic*, Book IV, paragraphs 440–443; see especially 443e. This ideal is equivalent, for Plato, to that of the just man, and also (see paragraph 444) to the healthy one.

7. Owen Barfield, *Saving the Appearances*, pp. 79–80; Robert Ornstein, *The Psychology of Consciousness* (Baltimore: Penguin Books, 1975), pp. 138, 183.

8. The discussion below follows that given by Michel Foucault in *The Order of Things* (New York: Vintage Books, 1973; orig. French edition 1966), chap. 2.

9. The first book of Agrippa's work has been translated into English as *The Philosophy of Natural Magic*, ed. L. W. de Laurence (Mokelumne Hill, Calif.: Health Research, 1972 reprint of 1913 ed.). The following quotations can be found on pp. 65, 71, 73, 77, 114, 210. Reprinted by permission of Health Research, Box 70, Mokelumne Hill, California 95245).

10. Cf. the similarity of the French words *aimant* (magnet) and *amant(e)* (lover.)

11. This theme is elaborated upon in Keith Thomas, *Religion and the Decline of Magic* (Harmondsworth: Penguin Books, 1973), and D. P.

Walker, *Spiritual and Demonic Magic from Ficino to Campanella* (London: The Warburg Institute, 1958).

12. On much of the following see Foucault, *The Order of Things,* chap. 3. Part I of *Don Quixote* appeared in 1605.

13. Owen Barfield, *Saving the Appearances,* pp. 32n, 42. Much of the argument given below follows the analysis developed in this work.

14. Stated in this way, the "common-sense" position that phenomena are wholly independent of consciousness and always have been appears to be so silly as to hardly warrant further comment. It *is,* however, the common-sense position, as well as the basic premise of all intellectual history or history of consciousness. Jaynes's study of human consciousness (see note 4 of this chapter) is so squarely based on this premise that he is ultimately forced to condemn every form of participating knowledge (poetry, music, art) as deluded and atavistic, and to champion the alienated intellect as the only reliable form of knowing (even though he comes to question that form, his own work included, by the end of the book). The Platonic ideal is thus taken to its ultimate psychotic conclusion. I should add that despite my criticism of this scientific ideal, I am in complete agreement with Barfield that a return to original participation is neither possible nor desirable at this point in human history.

15. From his book *De Vanitate,* and quoted by Carl Jung in *The Collected Works of C. G. Jung* (hereafter *CW*), trans. R. F. C. Hull, 2d ed. (Princeton: Princeton University Press, 1961–79), vol. 14, p. 35. See also Wolf Dieter Müller-Jahncke, "The Attitude of Agrippa von Nettesheim (1486–1535) Towards Alchemy," *Ambix* 22 (1975), 134-50. In England especially, alchemy was often seen as a con game, and the alchemical quest as akin to gambling fever. Chaucer ridiculed it in the "Canon Yeoman's Tale" as a waste of time and money.

16. Jung, *CW* 12, pt. 2. See also the fine collection in S. K. De Rola, *Alchemy: The Secret Art* (New York: Avon Books, 1973).

17. My description of Jung's work in this chapter is based on his *CW,* 12, 14, 15, and *Memories, Dreams, Reflections,* ed. Aniela Jaffé, trans. Richard and Clara Winston, rev. ed. (New York: Vintage Books, 1965); Anthony Storr, *Jung* (London: Fontana, 1973); Harold Stone, Prologue to Dora M. Kalff, *Sandplay* (San Francisco: Browser Press, 1971); and B. J. T. Dobbs, *The Foundations of Newton's Alchemy* (Cambridge: Cambridge University Press, 1975), pp. 26–34.

The word "gibberish," ironically enough, was first applied to the language of alchemy by outsiders, and was taken from Geber, the name of a thirteenth-century Italian or Catalan writer on the subject who in turn took his name from that of the eighth-century Arabian alchemist Jâbir ibn Hayyân.

18. I am following the terminology used by N. O. Brown in *Life Against Death* (Middletown, Conn.: Wesleyan University Press, 1970; orig. publ.

1959). For an interesting discussion of the language of dreams, see Ann Faraday, *The Dream Game* (New York: Perennial Library, 1976), pp. 54–57.

19. *CW* 12, pt. 2.

20. Brown, *Life Against Death*, p. 316.

21. Of course, barbarism is hardly the prerogative of modern man, although its scale probably is. Jung would conceivably have argued that the creation of a technology necessary to effect the genocide of modern times was itself part of the process of psychic repression.

22. "Sol et eius umbra perficiunt opus," from a work of 1618, quoted by Dobbs in her study of Newton, p. 31 and n.

23. The experience has an allegory in "The Fisherman and the Genie" from the *Arabian Nights*, in which the genie, once released from the bottle, threatens to kill the fisherman and is not easily persuaded to return whence he came. The Western version of this, naturally enough, deals with technology, and is captured in Mary Wollstonecraft Shelley's gothic novel, *Frankenstein*.

24. Actually, this is an incorrect correlate. Psychosis is the attempt to *salvage* the soul; it is only from the viewpoint of Western clinical psychiatry that it is regarded as purely negative. See the concluding pages of Chapter 4.

25. R. D. Laing, *The Politics of Experience* (New York: Ballantine Books, 1968).

26. R. D. Laing, *The Divided Self* (Harmondsworth, Penguin Books, 1965; orig. publ. 1959), pp. 200, 204–5.

27. Mircea Eliade, *The Forge and the Crucible*, trans. Stephen Corrin (New York: Harper Torchbooks, 1971; orig. French edition 1956), pp. 7–9, 30–33, 42, 54–57, 101–2.

28. Titus Burckhardt, *Alchemy*, trans. William Stoddart (Baltimore: Penguin Books, 1971; orig. German edition 1960), p. 25. The discussion of alchemical procedure that follows is taken from this book.

29. According to Frank Manuel, *A Portrait of Isaac Newton* (Cambridge: Harvard University Press, Belknap Press, 1968), p. 171, there were a total of twelve basic procedures, which he lists (following the system of Sir George Ripley, 1591) as: calcination, dissolution, separation, conjunction, putrefaction, congelation, cibation, sublimation, fermentation, exaltation, multiplication, and projection. The progress of the work was also charted by the various colors that were produced in the vessel, a type of index with obvious metallurgical roots. The "descent into chaos" of the initial solution was characterized by a blackening, or *nigredo*, followed by a bleaching, or *albedo*, and ending up (if all went well) with a reddening, or *rubedo*. But there was a whole series of intermediate colors also, and hence the term *cauda pavonis*, or peacock's tail, is frequently employed in the texts. Mercury produced a blackening, sulfur a reddening.

30. One of the best statements of the alchemical model of the human

personality, although it does not refer to alchemy as such, is Luke Rhinehart's hilarious novel, *The Dice Man* (London: Talmy, Franklin Ltd., 1971). The religious and psychoanalytic interpretation can be found in various sources, but I chose to paraphrase the interpretation provided by James Hillman in a lecture given in San Francisco on 11 December 1976. Hillman is editor of *Spring* and author of a number of works on Jungian psychology. A similar analysis of the nature of personality may be found in Hermann Hesse's brilliant novel *Steppenwolf*.

The quotation from Laing is on p. 190 of *The Politics of Experience and the Bird of Paradise* (Penguin Books, 1967). This book, like *The Divided Self*, is a profoundly alchemical work.

31. On Perry's work, see his book, *The Far Side of Madness* (Englewood Cliffs, N.J.: Prentice-Hall, 1974). The parallels between insanity and alchemy, and premodern thought in general, are discussed briefly at the end of Chapter 4 of the present work.

32. F. Sherwood Taylor, *The Alchemists* (New York: Henry Schuman, 1949), pp. 179–89. The testimony of Spinoza occurs in a letter he wrote to Jarrig Jellis in March 1667, reprinted in his *Posthumous Works*.

33. On alchemy as the key to nature see the various authors quoted by A. G. Debus in "Renaissance Chemistry and the Work of Robert Fludd," *Ambix* 14 (1967), 42–59. Agrippa discussed the relations between alchemy and numerous craft processes (see the article by Müller-Jahncke cited in note 15 of this chap.); and its relationship to mining, metallurgy, and pottery is discussed at length by Eliade in *The Forge and the Crucible*. The relationship between alchemy and medicine is the subject of a large literature, and has been explored in the work of Paracelsus and his followers by Allen Debus and Walter Pagel. Finally, alchemy as a yoga has been discussed by Eliade, Jung, and a host of other writers. Of particular interest are Burckhardt, *Alchemy*, and Maurice Aniane, "Notes sur l'alchimie, 'Yoga' cosmologique de la chrétienté médiévale," in Jacques Masui, ed., *Yoga, science de l'homme intégral* (Paris: Cahiers du Sud, 1953), pp. 243–73.

34. Chinua Achebe, *Things Fall Apart* (New York: Fawcett World Library, 1959). The first three books of the Castaneda tetralogy, *The Teachings of Don Juan*, *A Separate Reality*, and *Journey to Ixtlan*, deal with the animistic world view mostly from the inside. The fourth, *Tales of Power*, spells out the epistemology of sorcery in exact detail.

35. Reprinted by permission of G. P. Putnam's Sons from *Seeing Castaneda*, edited by Daniel Noel, p. 53. Copyright © 1976 by Daniel Noel.

36. Philip Wheelwright, ed., *The Presocratics* (New York: The Odyssey Press, 1966), p. 52.

37. Taylor, *The Alchemists*, pp. 233–34.

38. See Chapter 4 for a discussion of Newton's alchemy. Something of a cottage industry has developed among historians of science regarding

Newton as an alchemist, and there is by now a good bit of literature on the subject. The interested reader might wish to consult any of the following: Frank Manuel, *A Portrait of Isaac Newton;* J. E. McGuire and P. M. Rattansi, "Newton and the 'Pipes of Pan'," *Notes and Records of the Royal Society of London* 21 (1966), 108–43; Betty Dobbs, *The Foundations of Newton's Alchemy;* R. S. Westfall, "The Role of Alchemy in Newton's Career," in M. L. R. Bonelli and W. R. Shea, eds., *Reason, Experiment, and Mysticism in the Scientific Revolution* (New York: Science History Publications, 1975), pp. 189–232, and also his "Newton and the Hermetic Tradition," in A. G. Debus, ed., *Science, Medicine and Society in the Renaissance,* 2 vols. (New York: Neale Watson, 1972), 2, 183–98; P. M. Rattansi, "Newton's Alchemical Studies," in the Debus volume, pp. 167–98; and the remarkable essay by David Kubrin, "Newton's Inside Out! Magic, Class Struggle, and the Rise of Mechanism in the West," in Harry Woolf, ed., *The Analytic Spirit* (Ithaca, N.Y.: Cornell University Press, 1981).

Christopher Hill provides a brilliant discussion of seventeenth-century radical ideas, including those of the occult sciences, in *The World Turned Upside Down* (New York: Viking, 1972).

39. There is, furthermore, a still existing underground of practical alchemists. See Jacques Sadoul, *Alchemists and Gold* (London: Neville Spearman, 1972), and Armand Barbault, *Gold of a Thousand Mornings,* trans. Robin Campbell (London: Neville Spearman, 1975).

40. Both quotations from Magritte can be found in Eddie Wolfram's Introduction to David Larkin, ed., *Magritte* (New York: Ballantine Books, 1972). The link between alchemy and surrealism is mentioned briefly by E. R. Chamberlin in *Everyday Life in Renaissance Times* (New York: Capricorn Books, 1965), p. 175.

41. See Walker, *Spiritual and Demonic Magic.*

42. Eliade, *Forge and Crucible,* pp. 172–73. Cf. Brown, *Life Against Death,* p. 258.

43. Paolo Rossi, *Philosophy, Technology and the Arts in the Early Modern Era,* trans. Salvator Attanasio (New York: Harper Torchbooks, 1970; orig. Italian edition 1962), p. 28. The idea that Hermeticism was a major factor in the rise of the experimental method is now accepted by many historians. In addition to Rossi, several of the authors cited in note 38 talk in these terms, as does Eliade in *Forge and Crucible,* Frances A. Yates in *Giordano Bruno and the Hermetic Tradition* (New York: Vintage Books, 1969), and Christopher Hill in *Intellectual Origins of the English Revolution* (London: Panther Books, 1972; orig. publ. 1965). See also the Introduction in A. G. Debus, ed., John Dee, *The Mathematicall Preface to the Elements of Geometrie of Euclid of Megara,* 1570 (New York: Science History Publications, 1975).

However, Robert S. Westman has seriously questioned the thesis, and J. E. McGuire has significantly distanced himself from his earlier position, in essays published under the title of *Hermeticism and the Scientific Revolution* (Los Angeles: William Andrews Clark Memorial Library, 1977).

44. Keith Thomas, *Religion and the Decline of Magic.*

45. Yates, *Giordano Bruno,* p. 99.

46. *Elim* was also an allusion to a biblical place name, mentioned in Exodus 15:27 and 16:1. For a biographical sketch of Delmedigo see the *Encyclopaedia Judaica,* 5 (1972), 1478–82. The plate from Fludd occurs in the second volume of his book, *Utriusque cosmi maioris scilicet et minoris metaphysica, physica atque technica historia, in duo secundum cosmi differentiam divisa.*

47. Rossi, *Philosophy, Technology and the Arts,* p. 149.

48. On Dee see Peter J. French, *John Dee* (London: Routledge & Kegan Paul, 1972), and the work by Debus cited above, note 43. On Campanella see Yates, *Giordano Bruno,* passim. The "blurring" of magic and technology can be seen in Agrippa's *De Occulta Philosophia.*

49. Quoted in Hill, *Intellectual Origins,* p. 149.

50. On Ficino's astrology, and Bacon's reaction, see Walker, *Spiritual and Demonic Magic.*

51. Erwin F. Lange, "Alchemy and the Sixteenth Century Metallurgists," *Ambix* 13 (1966), 92–95. Apparently the first of this tradition, the *Bergbüchlein* of 1505, contained an equal mixture of the metallurgical and the alchemical (see the discussion in Eliade, *Forge and Crucible,* pp. 47–49). Biringuccio's work of only thirty-five years later denounced alchemy, although according to Rossi, *Philosophy, Technology and the Arts,* p. 52n, he was uncertain about his own opinion on the subject. The first edition of Agricola appeared (without illustrations) in 1546, and he was definitely not confused about his attitudes toward alchemy.

52. Quoted in Rossi, *Philosophy, Technology and the Arts,* p. 71.

53. Ibid., pp. 43–55, and the Preface to *De Re Metallica,* trans. Herbert Clark Hoover and Lou Henry Hoover (New York: Dover Publications, 1950, orig. English trans. 1912).

54. Thomas, *Religion and the Decline of Magic,* chap. 2.

55. The following discussion is based on Jung, *CW* 12 and 14.

56. Dobbs, *Foundations of Newton's Alchemy,* pp. 34–36.

57. The usual symbol for Christ used in this way was the unicorn, and this can be seen, for example, in the famous unicorn tapestry cycle on display at the Cloisters in upper Manhattan.

58. The discussion below is based on the following sources: Richard H. Popkin, "Father Mersenne's War Against Pyrrhonism," *The Modern Schoolman* 24 (1957), 61–78; A. R. Hall, *The Scientific Revolution* (Boston: Beacon Press, 1956), pp. 196–97; Robert H. Kargon, *Atomism in England from Hariot to Newton* (Oxford: Clarendon Press, 1966); Michael Maier, *Laws of the Fraternity of the Rosie Cross* (Los Angeles: The Philosophical

Research Society, 1976, from the English edition of 1656; orig. Latin edition 1618); A. G. Debus, "Renaissance Alchemy and the Work of Robert Fludd," *The English Paracelsians* (London: Oldbourne Book Co., 1965), and "The Chemical Debates of the Seventeenth Century: The Reaction to Robert Fludd and Jean Baptiste van Helmont," in M. L. R. Bonelli and W. R. Shea, *Reason, Experiment, and Mysticism*, pp. 19–47; and Dobbs, *The Foundations of Newton's Alchemy*, pp. 53–63. Also useful are Robert Lenoble, *Mersenne ou la naissance du mécanisme* (Paris: Librairie Philosophique J. Vrin, 1943), and Francis A. Yates, *The Rosicrucian Enlightenment* (London: Routledge & Kegan Paul, 1972).

The French attempt to establish a stable world philosophy based on mechanism, and directly opposed to the dialectical principles of Hermeticism, occurred in the context of growing political absolutism and peasant rebellion, the latter especially frequent from 1623 to 1648. This theme is explored by Carolyn Merchant in *The Death of Nature* (New York: Harper & Row, 1980), chapter 8, and I am grateful to her for allowing me to read the manuscript version of this part of her work. My own discussion deals primarily with the religious aspects of the attack on Hermeticism, but the reader should be aware that church issues were not separate from issues of state in the minds of the protagonists. Thus my own discussion necessarily follows the line of reasoning developed by Professor Merchant.

59. In the context of the time, as Robert Kargon points out in *Atomism in England*, there were significant differences between the various atomists and corpuscularians. Gassendi's idea was that motion was essential to matter, bestowed on it by God at the creation. Hence, his system was based on the views of the ancient atomist Epicurus, but heavily Christianized so as to be acceptable. From the vantage point of the late seventeenth century and after, however, Descartes, Hobbes, and Gassendi had all formulated an impact physics.

60. A more rational debate than the attack on alchemy, however, was the one between Fludd and Johannes Kepler, which also weakened alchemy publicly and helped to establish the fact-value distinction. Nevertheless, I do not think this debate, which came just before the attack by Mersenne and Gassendi, can be seen apart from the rise of the technological tradition and the religious developments described above. Kepler certainly was (despite his own very extensive Hermeticism) arguing for an empirical, rather than an allegorical, view of the cosmos; but the "conditions enabling that system to be thought" (as Foucault puts it) lay in the exoteric–esoteric split that had been building for more than a century before the debate took place. What we call empiricism, which by definition is an exclusion of occult causes, is precisely the product of the changes described in this chapter.

An interesting discussion of the Kepler-Fludd debate may be found in W. Pauli, "The Influence of Archetypal Ideas on the Scientific Theories of

Kepler," in C. G. Jung and W. Pauli, *The Interpretation of Nature and the Psyche*, trans. Priscilla Silz (London: Routledge & Kegan Paul, 1955), pp. 151–240.

61. Thomas, *Religion and the Decline of Magic*, chap. 3.

62. Ibid., p. 130.

63. Manuel, *Portrait of Isaac Newton*, pp. 59, 380.

64. Hill, *World Turned Upside Down*, p. 262.

CHAPTER 4. *The Disenchantment of the World (2)*

1. Actually, Newton's interest in alchemy was revealed soon after his death, but as noted below, in the context of eighteenth-century rationalism the priority was to "clear" him of any "charges" of having been an alchemist. L. T. More apparently neglected, or did not have access to, Newton's alchemical and theological manuscripts when he wrote *Isaac Newton: A Biography* (London: Constable, 1934), and thus did not have to trouble himself too much about integrating the rational and the mystical aspects of the man (a dichotomy which, I hope to show, is spurious in any event).

2. Quoted in B. J. T. Dobbs, *The Foundations of Newton's Alchemy* (Cambridge: Cambridge University Press, 1975), pp. 13–14.

3. Frank E. Manuel, *A Portrait of Isaac Newton* (Cambridge: Harvard University Press, Belknap Press, 1968). For Kubrin's study see Harry Woolf, ed., *The Analytic Spirit* (Ithaca, N.Y.: Cornell University Press, 1981). Kubrin's essay is given fuller treatment in an earlier work of his, *How Sir Isaac Newton Helped Restore Law 'n' Order to the West* (Privately printed, 1972), copies of which are on deposit in the Library of Congress.

4. The following sketch is taken from Manuel, *Portrait of Isaac Newton*, pp. 23–67. Manuel's model is based on the work of Erik Erikson, who sees in all leading figures of the age (his own studies were of Luther and Ghandi) extreme expressions of trends already present throughout the populace. Manuel was able to develop this theme well in Newton's case because of the existence of four adolescent notebooks that reflect the severe repression and depression of the Puritan mentality.

On anxiety reactions see N. O. Brown, *Life Against Death* (Middletown, Conn.: Wesleyan University Press, 1970; orig. publ. 1959), esp. pp. 114ff.; John Bowlby, *Separation* (New York: Basic Books, 1973); and Erikson's pioneering work, *Childhood and Society*, 2d ed., rev. and enl. (New York: Norton, 1963).

5. Manuel, *Portrait of Isaac Newton*, p. 380.

6. Géza Róheim, *Magic and Schizophrenia* (Bloomington: Indiana University Press, 1970; orig. publ. 1955).

7. D. P. Walker, *The Ancient Theology. Studies in Christian Platonism from the Fifteenth to the Eighteenth Century* (London: Gerald Duckworth, 1972).

8. Cited by Rollo May in John Brockman, ed., *About Bateson* (New York: Dutton, 1977), p. 91.

9. This emerges in an important theme in Betty Dobbs's study, *The Foundations of Newton's Alchemy.*

10. Kubrin, "Newton's Inside Out!" The discussion that follows relies heavily on this essay, and I am very grateful to Mr. Kubrin for allowing me to read the unpublished version. On the volume of alchemical publications see also Keith Thomas, *Religion and the Decline of Magic* (Harmondsworth: Penguin Books, 1973), p. 270.

11. See works such as *The Century of Revolution, God's Englishman,* and especially *The World Turned Upside Down* (New York: Viking, 1972).

12. See, for example, the revelation recorded by the Ranter Abiezer Coppe, reprinted in Norman Cohn, *The Pursuit of the Millennium* (London: Paladin, 1970; orig. publ. 1957), pp. 319–30. Cohn is horrified by such a text, but one's attitude clearly depends on whether one is inside or outside the experience.

13. Thomas, *Religion and the Decline of Magic,* p. 322.

14. Cf. the concluding pages of Chapter 3. Note that I am using the term "middle class" here in the traditional Marxist sense, that is (in the English case), to refer to the economic and political interests that opposed the king, not in the modern sociological sense of group identification, socioeconomic stratification, and so on.

15. Some of the radical leaders/occultists include William Lilly, John Everard, Lawrence Clarkson, Nicholas Culpepper, Gerard Winstanley, William Dell, John Webster, John Allin, and Thomas Tryon. Statements by clerics may be found in P. M. Rattansi, "Paracelsus and the Puritan Revolution," *Ambix* 11 (1963), 24–32.

16. Hill, *World Turned Upside Down,* pp. 144, 238, 287.

17. The quotes from Newton are cited in Kubrin, "Newton's Inside Out!" For alchemical language in Newton, see H. S. Thayer, ed., *Newton's Philosophy of Nature* (New York: Hafner, 1953), pp. 49, 84–91, 164–65.

18. R. S. Westfall, "The Role of Alchemy in Newton's Career," in M. L. R. Bonelli and W. R. Shea, eds., *Reason, Experiment, and Mysticism in the Scientific Revolution* (New York: Science History Publications, 1975), pp. 189–232.

19. See Newton's *Chronology of Ancient Kingdoms Amended . . .* (London, 1728), esp. pp. 332–46, and *A Dissertation upon the Sacred Cubit of the Jews,* in John Greaves, *Miscellaneous Works . . .* (London, 1737), vol. 2.

20. On this see also Margaret C. Jacob, *The Newtonians and the English Revolution, 1689–1720* (Ithaca, N.Y.: Cornell University Press, 1977).

21. I. B. Cohen and Alexandre Koyré, "The Case of the Missing *Tanquam:* Leibniz, Newton, and Clarke," *Isis* 52 (1961), 555–67.

22. For a contemporary view of the earth as alive see Lewis Thomas, *The Lives of a Cell* (New York: Viking, 1974).

23. E. P. Thompson, *The Making of the English Working Class* (New York: Pantheon, 1964), esp. chap. 11.

24. Quoted in Brown, *Life Against Death,* p. 108.

25. The same hardening can be seen in paintings of a later president of the Royal Society, Humphry Davy, in Plates 11 and 12 of my *Social Change and Scientific Organization* (London and Ithaca, N.Y.: Heinemann Educational Books and Cornell University Press, 1978), and should be compared to Plates 24 and 25, which juxtapose portraits of the young and old Michael Faraday. As I discuss in that work, Faraday was a religious mystic and something of a closet Hermeticist, believing that matter was essentially spiritual in nature. The photograph of Faraday as an older man is remarkable for its childlike nature: the gentle expression and the bright, almost glowing eyes.

26. Quoted in Hill, *World Turned Upside Down,* p. 287.

27. David V. Erdman, ed., *The Poetry and Prose of William Blake* (Garden City, N.Y.: Doubleday, 1965), p. 693.

28. Milton Klonsky, *William Blake: The Seer and His Visions* (New York: Harmony Books, 1977), p. 62.

29. Hill, *World Turned Upside Down,* p. 311.

30. Ibid., p. 236.

31. The following discussion is based (partly) on R. D. Laing, *The Divided Self* (Harmondsworth: Penguin Books, 1965; orig. publ. 1959), esp. pp. 140–41, 148, 151, 179, 198.

CHAPTER 5. *Prolegomena to Any Future Metaphysics*

1. The title of this chapter is taken from that of a book of the same name published by Immanuel Kant in 1783, two years after the first edition of his famous *Critique of Pure Reason.* I am not a Kantian and this chapter is not an attempt at Kantian analysis. Nevertheless, my own work does attempt to emulate Kant in the following ways, and hence I did not feel I could do better than to use a Kantian title most appropriate to my own goals:

(a) Kant made an attempt to state what he believed were the central problems of philosophy during his own day, and to distill principles that he hoped would be valid for all human knowledge.

(b) Kant realized that any future metaphysics must have a prolegomena, that is, some sort of preface setting out what the criteria of a new science might be.

(c) Kant was perhaps the first Western philosopher in the modern period to recognize that the mind is not simply bombarded by sense impressions, but actually plays a role in shaping what it perceives.

2. Quoted in N. O. Brown, *Life Against Death* (Middletown, Conn.: Wesleyan University Press, 1970; orig. publ. 1959), p. 315.

Notes

3. Michael Polanyi, *Personal Knowledge*, corrected ed. (Chicago: University of Chicago Press, 1962); Owen Barfield, *Saving the Appearances* (New York: Harcourt, Brace & World, 1965).

4. Polanyi, *Personal Knowledge*, p. 294.

5. It should be added that in these illustrations it probably is possible for an observer to see both images simultaneously if he or she is in a meditative or "alpha" brain-wave state. Under normal conditions, however, the brain selects one over the other.

6. See Polanyi, *Personal Knowledge*, pp. 69–131, 249–61, and passim; see also pp. 49–65. The specific issue of language acquisition is discussed by Daniel Yankelovich and William Barrett (drawing on Noam Chomsky) in *Ego and Instinct* (New York: Vintage Books, 1971), pp. 388–92, and by Susanne Langer in *Philosophy in a New Key*, 3d ed. (Cambridge: Harvard University Press, 1957), pp. 122–23, 122n.

7. From page 101 of *Personal Knowledge* by Michael Polanyi, copyright © 1958, 1962; reprinted by permission of the University of Chicago Press.

8. Ibid., pp. 60–70, 88–90, 123, 162.

9. The following discussion is taken from Barfield, *Saving the Appearances*, pp. 24–25, 32n, 40, 43, 81, and passim. What Barfield calls "alpha-thinking" (see below) is not to be confused with the generation of alpha brain waves in altered states of consciousness (above, note 5). Barfield's "alpha-thinking" is actually a type of "beta-thinking," in the jargon of recent brain research.

10. Retaining what has been called the "illusion of the first time" is quite difficult once you become skilled at an activity. It is this sense of wonder that adults most envy in very young children.

11. Peter Achinstein, *Concepts of Science* (Baltimore: The Johns Hopkins University Press, 1968), p. 164. The above example is only touched on in this book. I had the good fortune to be a student of Professor Achinstein during my graduate years, and have elaborated the example given in his book into the much fuller version that he provided in the classroom.

Alan Watts's favorite example of confusing map with territory was sitting down in a restaurant and eating the menu instead of the dinner, an act that he saw as a metaphor for modern society in general.

12. The best one-volume discussion of the subject for the layman, and it is not easy going, is *The Strange Story of the Quantum*, by Banesh Hoffman, rev. ed. (Harmondsworth: Penguin Books, 1959). I have also found Jeremy Bernstein's *Einstein* (London: Fontana, 1973), and Werner Heisenberg's *Physics and Philosophy* (New York: Harper Torchbooks, 1962), helpful in understanding the subject.

When I state that the scientific establishment pretends that quantum mechanics does not exist, I mean this in the philosophical rather than the literal sense. Quantum mechanics is certainly recognized as a legitimate area of research, and a recent article in *Scientific American* by Bernard d'Espagnat ("The Quantum Theory and Reality," 241 [November 1979], 158–81), does not mince words as to how epistemologically radical the subject truly is. But virtually all scientists proceed with their work as though they were detached observers, and the traditional subject/object dichotomy is embedded in the curricula and textbooks of all high school and college science teaching.

Some of the most advanced work using quantum mechanics to create a new scientific metaphysics is being done by David Finkelstein of Yeshiva University. See, for example, his articles on the "Space-Time Code" in *Physical Review* 184 (25 August 1969), 1261–70; and *Physical Review D* 5 (15 January 1972), 320–28, (15 June 1972), 2922–31, and 9 (15 April 1974), 2219–31. Finkelstein also has an interesting paper on "Matter, Space and Logic" in *Boston Studies in the Philosophy of Science* 5 (1969), 199–215.

13. See Northrup's introduction in Heisenberg, *Physics and Philosophy*, pp. 6–10. The quotations from Heisenberg below are taken from this book, pp. 29, 41, 58. See also pp. 81, 130, 144.

According to Norwood Russell Hanson, if one were to argue that the uncertainty relations do not mean the electrons actually lack a simultaneous position and momentum, one would be essentially arguing that electrons *are* in precisely defined states but that we cannot define them because of crude techniques of investigation. This valiant attempt to save classical notions of reality will not work. As Hanson points out, this position "seeks what no physical theory can hope for—a knowledge of nature that transcends what our best hypotheses and experiments suggest." The close connection between epistemology and ontology becomes obvious here. If we cannot know an object in the classical Cartesian sense, how can we argue that it conforms to classical notions of reality? Arguing that it must conform to the usual subject/object relations turns the Cartesian paradigm into a faith, not a science; which is what it always was anyway.

See N. R. Hanson, "Quantum Mechanics, Philosophical Implications of," in Paul Edwards, ed., *The Encyclopedia of Philosophy* (New York: Macmillan, 1967), 7:44.

14. This attempt to find the ultimate material entity is still, foolishly, going on. Of the two hundred or so nuclear particles now recognized as existing, 90 percent of these have been discovered in the postwar era, suggesting that reality is more a function of the national budget than anything else. Since 1964, atomic physicists have posited the existence of "quarks" (a word taken from *Finnegan's Wake*) to explain these particles, but their number has multiplied to the point that we may soon have a quark to explain each particle. Nor is this the end: to explain quarks, "hidden variables" have now been suggested. In fact, there is no end to this process. As Geoffrey Chew has pointed out, we detect particles because they interact with the observer, but in order to do so they must have some internal structure. This means that we can in principle never get to

some object that has no internal structure, for a truly elementary particle could not be subject to any forces that would allow us to detect its existence (if we find it by its weight, for example, then it must contain something within it producing a gravitational field). On the Cartesian model we shall be chasing "hidden variables" to the end of time. The disarray in modern physics became embarrassingly clear at the 1978 meeting of the American Physical Society in San Francisco, at which an appeal was made for a new Einstein to sort things out. The cul-de-sac of Cartesianism came out in a remark made by one Berkeley physicist, that although no one knew what the proliferation of particles meant, at least we could measure them with great precision (!). On a more intelligent level, Werner Heisenberg called for an end to the concept of the elementary particle in 1975. William Irwin Thompson's remark that an "elementary particle is what happens when you build an accelerator" is not without relevance here.

See Fritjof Capra, *The Tao of Physics* (Berkeley: Shambhala, 1975), pp. 273–74; "Scientist's Call for Another Einstein," *San Francisco Chronicle*, 24 January 1978; "Monitor," *New Scientist*, 24 July 1975, p. 196; and William Irwin Thompson, "Notes on an Emerging Planet," in Michael Katz et al., eds., *Earth's Answer* (New York: Harper & Row, 1977), p. 210.

15. H. Forwald, *Mind, Matter and Gravitation* (New York: Parapsychology Foundation, 1969). Forwald, a retired engineer and inventor, performed these experiments over a period of two decades.

16. For example, Capra, *The Tao of Physics;* Lawrence LeShan, *The Medium, the Mystic, and the Physicist* (New York: Ballantine Books, 1975; orig. publ. 1966); Gary Zukav, *The Dancing Wu Li Masters* (New York: William Morrow, 1979).

17. E. H. Walker, "Consciousness in the Quantum Theory of Measurement," *Journal for the Study of Consciousness* 5 (1972), Part 1, no. 1, p. 46; Part 2, no. 2, p. 257; "The Nature of Consciousness," *Mathematical Biosciences* 7 (1970), 175.

18. Yankelovich and Barrett, *Ego and Instinct*, p. 203.

19. Gregory Bateson, *Steps to an Ecology of Mind* (London: Paladin, 1973; New York: Ballantine, 1972), p. 436 British edition, p. 461 American edition. The two modalities of human awareness are called "tonal" and "nagual" in some anthropological literature, and an excellent explication of their relationship may be found in the second half of Carlos Castaneda, *Tales of Power* (New York: Simon and Schuster, 1974). As in the case of Bateson's work, Casteneda's provides a brilliant model of holistic knowing. Unlike Bateson's work, it stops at the point that the model is delineated.

20. This recognition reflects perfectly the *internal* osmosis that goes on in holistic consciousness between the conscious and unconscious mind (nucleus and cell). In such consciousness the barrier between the two modalities disintegrates; they interpenetrate and become more like each

other. This process is accompanied by an external alteration in which Self and Other are not seen as so sharply distinguished.

21. Hanson, "Quantum Mechanics," p. 46.

22. Gregory Bateson, "Style, Grace and Information in Primitive Art," in *Steps to an Ecology of Mind*, p. 109n British edition, and p. 136n American edition.

23. Quoted in Arthur Koestler, *The Roots of Coincidence* (New York: Random House, 1972), p. 55.

24. Peter Koestenbaum, *Managing Anxiety* (Englewood Cliffs, N.J.: Prentice-Hall, 1974), pp. 11–13.

25. Brown, *Life Against Death*, pp. 94-5, 273-4. Both Freud and Reich made this point as well, at least by way of analogy. Cf. Wilhelm Reich, *The Function of the Orgasm*, trans. Vincent R. Carfagno (New York: Pocket Books, 1975; orig. German ed. 1942), pp. 33, 283.

26. Barfield, *Saving the Appearances*, pp. 136, 144, 160.

27. All terms that make this distinction between inner and outer, thus perpetuating mind/body, subject/object dualism, should be put in inverted commas. In this category I would include phrases such as "phenomena," "data," "the given," and so on. We need a new vocabulary that reinforces the ecological sense of reality.

28. At the risk of belaboring a point, I am not suggesting, as Berkeley did, that events would not exist were we absent, but only that the nature of what is going on is in some way dependent upon our participation in the events. What occurs in our absence would thus be irrelevant.

As for modern cosmology, the latest word, from the Lick Observatory of the University of California, is that the universe is actually collapsing. Or rather, it will apparently expand for another twenty billion years, and then collapse for the next thirty billion after that. Once again, the whole thing seems to resonate with the sociology of knowledge. As Europe began expand its geographic and economic horizons, the universe went from completely closed to infinitely open. Now that the futures of science, technology, linear progress, and industrial society have all become rather questionable, the cosmos has curiously begun to contract! See "New Evidence Backs A Collapsing Universe," *San Francisco Chronicle*, 30 June 1978.

29. Polanyi, *Personal Knowledge*, pp. 288–94. An elaboration of the circularity of modern science can also be found in Max Marwick, "Is science a form of witchcraft?" *New Scientist*, 5 September 1974, pp. 578–81.

30. The sociology of knowledge did exist prior to modern times, but not in a serious or systematic way. Protagoras states that "man is the measure of all things," but he is referring to what an individual believes, not a culture, and he makes no mention of social influences. Plato says at one point that the lower classes cannot know the truth because their work distorts their minds and bodies; but this statement is really a sociology of error rather than an examination of the social roots of an epistemology (al-

though it must be admitted that the line between these two is not altogether clear). Though there is a subdued theme in Plato that social circumstances shape the subject of knowing, it is much overpowered by the notion of the immutability of the Forms, and it is not developed as an ongoing critique in any event. The subject does not get any rigorous attention until the Enlightenment, and the sociology of knowledge does not constitute a serious discipline prior to Marx's classic formulation of the relationship between existence and consciousness. (On this point see Werner Stark, "Sociology of Knowledge," in Paul Edwards, ed., *The Encyclopedia of Philosophy*, 7: 475–78.)

The paradoxes that the discipline is capable of generating, however, were known as far back as the fifth century B.C. Thus, what is called "Mannheim's Paradox" is a version of the ancient puzzle known as the "Liar's Paradox" (A Greek said, "All Greeks are liars." Was he telling the truth?). In other words, if one takes Mannheim seriously, his argument that knowledge is situation-bound must apply to that argument itself ("What sort of culture produced the sociology of knowledge?"). But if it does apply, then the argument is wrong, or at least thrown in doubt; and if the argument is wrong (knowledge is *not* situation-bound), then it might be right, and so on. (Plato uses the same line of reasoning against the doctrine of Protagoras in the *Theaetetus*, 171A.) Various Greek schools of thought, such as those of the Megarians and the Eleatics, delighted in elaborating puzzles of this sort; and in a more serious vein, the so-called Third Man Argument of Plato's dialogue *Parmenides* presents the paradox of infinite regress as a threat to the theory of Forms. But we should be clear that although these various pardoxes do involve radical relativism in suggesting that there may be no fixed truth, they are strictly problems of logic, not equivalent to the sociology of knowledge. That is, they do not develop the theme that information about the world is relative because it is socially conditioned or culture-bound.

Finally, it is important to add that commentary itself is not an issue here; there is plenty of commentary and analysis in the Talmud, for example. But the rabbis of the Middle Ages did not, to my knowledge, analyze the nature of their own analysis, any more than cultures that lived by myth had myths about the general nature or epistemological status of mythology—that is, they had no myths explaining how myth per se ascertains truth.

31. Friedrich Nietzsche, *The Birth of Tragedy*, trans. Francis Golffing (Garden City, N.Y.: Doubleday Anchor, 1956), p. 95.

CHAPTER 6. *Eros Regained*

1. Susanne Langer, *Philosophy in a New Key*, 3d ed. (Cambridge: Harvard University Press, 1957), pp. 88, 92.

2. Erich Neumann, *The Child*, trans. Ralph Mannheim (New York: Harper & Row, 1976), pp. 11–17, 28, 30; Sam Keen, *Apology for Wonder* (New York: Harper & Row, 1969), p. 46. For two studies of child development that view the first few weeks of life in Freudian terms, see Margaret S. Mahler et al., *The Psychological Birth of the Human Infant: Symbiosis and Individuation* (New York: Basic Books, 1975), and Edith Jacobson, *The Self and the Object World* (New York: International Universities Press, 1964). Although Neumann described the first three months of life in these terms, he did take issue with the term "narcissism" as implying a type of power relationship, which is not possible if an other is totally unrecognized.

The letter from Rolland is mentioned in a footnote inserted in the 1931 edition of *Civilization and Its Discontents*. Freud admitted that the letter "caused me no small difficulty. I cannot discover this 'oceanic' feeling in myself. It is not easy to deal scientifically with feelings. One can attempt to describe their physiological signs." We can begin to understand why Freud's view of human life was so pessimistic. See James Strachey, ed., *The Standard Edition of the Complete Psychological Works of Sigmund Freud*, 24 vols. (London: The Hogarth Press, 1953–74), 21: 65 and n.

3. At least, this was Freud's position as of 1923, and before 1902. Between those two dates, Freud saw the ego itself as a set of instincts instead of a structure deriving its energy from the id. This position became the central plank of ego psychology, with Heinz Hartmann its leading exponent. For an excellent overview of the evolution of early psychoanalytic thought, see Daniel Yankelovich and William Barrett, *Ego and Instinct* (New York: Vintage Books, 1971), esp. pp. 25–114.

4. Cf. Gordon Rattray Taylor's intelligent discussion of "hard" versus "soft" ego in *Rethink* (Harmondsworth: Penguin Books, 1974), pp. 81–90, 109ff. What modern psychiatry calls "ego-strength" is more often really ego rigidity, and actually quite brittle. The equation of muted ego virtues with mental illness is characteristic of societies that define health in terms of productive capacity.

5. T. G. R. Bower, *The Perceptual World of the Child* (Cambridge: Harvard University Press, 1977), pp. 19–21, 28.

6. Ibid., pp. 34, 49–50; Mahler, *Psychological Birth of the Human Infant*, pp. 46–47, 52–56. I am personally skeptical of this time scale. Although (see below) perceptual development is not the same thing as ego-development, it is doubtful that the unspecific smile lasts for three months or that comparative scanning begins only at age seven months. Joseph Lichtenberg recently demonstrated that at the age of fourteen days, the neonate distinguishes between his mother's face and that of unknown female. See "New findings about the newborn," *San Francisco Examiner*, 28 May 1980.

7. Mahler, *Psychological Birth of the Human Infant*, p. 223n, and Maurice Merleau-Ponty, "The Child's Relations with Others," trans. William

Notes

Cobb, in James M. Edie, ed., *The Primacy of Perception* (Evanston, Ill.: Northwestern University Press, 1964), pp. 125–26. Merleau-Ponty's discussion is largely based on the work of the brilliant and relatively unknown Marxist child psychologist, Henri Wallon, which stands in sharp contrast to that of Piaget. As of this writing, Wallon would appear to be the only scientist who did extensive studies of children's behavior in front of the mirror, which Merleau-Ponty discusses on pp. 125–40 of his essay. (According to Mahler, such a research project will soon be published by John B. McDevitt.) For more on Wallon, see the Winter 1972/73 issue of the *International Journal of Mental Health,* as well as his article, "Comment se développe, chez l'enfant, la notion du corps propre," *Journal de Psychologie* (1931), 705–48.

8. Mahler, *Psychological Birth of the Human Infant,* pp. 67, 71, 77–92, 101; R. D. Laing, *The Divided Self* (Harmondsworth: Penguin Books, 1965; first publ. 1959), pp. 115–19.

9. Merleau-Ponty, "The Child's Relations with Others," pp. 136–37, 152–53.

10. Yankelovich and Barrett, *Ego and Instinct,* pp. 320, 386–92, 396–7. This point raises the problem of how language ever arose at all, which has never been solved. On this matter, and material on children raised by animals, see Langer, *Philosophy in a New Key,* pp. 108–42, and passim. Ashley Montagu presents a Darwinian theory of the origins of speech in *The Human Revolution* (New York: The World Publishing Company, 1965), pp. 108–13.

11. Bower, *Perceptual World of the Child,* p. 42.

12. Philippe Ariès, *Centuries of Childhood,* trans. Robert Baldick (New York: Vintage Books, 1962), pp. 103–6.

13. See, for example, Mahler, *Psychological Birth of the Human Infant,* p. 35. In general, I find this study highly teleological, with infants regarded almost as subhuman, but "redeemed" in that they are, after all, going to become adults. The authors do not seem to realize that the scientific terms used to describe childhood, which include "narcissism," "hallucinatory disorientation," and even "autism," are loaded and that such terms assume that the adult perception of the world is correct and anything else is incorrect.

The question of innate and acquired is discussed later on in this chapter. The importance of socialization was a major feature of Wallon's work (see above, note 7).

14. This material is taken from remarks made by John Kennell in the "General Discussion" section of Evelyn B. Thoman, ed., *Origins of the Infant's Social Responsiveness* (Hilldale, N.J.: Lawrence Erlbaum Associates, 1979), pp. 435–36.

15. Stuart A. Queen and Robert W. Habenstein, *The Family in Various Cultures,* 4th ed. (New York: Lippincott, 1974; first publ. 1952), p. 164; John

Ruhräh, *Pediatrics of the Past* (New York: Paul B. Hoeber, 1925), p. 34; and Ian G. Wickes, "A History of Infant Feeding," *Archives of Disease in Childhood* 28 (1953), 156.

"Extended" is a loaded term, since the time span is being measured from our point of view. It might be more correct to call the twentieth-century period of lactation "curtailed."

16. Ashley Montagu, *Touching: The Human Significance of the Skin*, 2d ed. (New York: Harper & Row, 1978), pp. 124, 187, 190, 199, 203, and chap. 7, passim.

17. The study of Bali is *Balinese Character* by Bateson and Mead, and is discussed by Montagu in *Touching*, pp. 115–18 (cf. Chapter 7, note 16). See also chap. 7 of Montagu on comparative cultural studies, and Beatrice B. Whiting, ed., *Six Cultures* (New York: John Wiley, 1963). On Ariès see note 12, above. The "new" books on childbirth and infantile sexuality include Alayne Yates, *Sex Without Shame* (New York: William Morrow, 1978); Frederick Leboyer, *Birth Without Violence* (New York: Knopf, 1975); and Fernand Lamaze, *Painless Childbirth* (New York: Pocket Books, 1977).

18. The following discussion is taken from Ariès, *Centuries of Childhood*, esp. pp. 10, 33–34, 52, 61, 107, 114–16, 254–60, 264, 353–56, 398–99, 405, 414–15. See also Lawrence Stone, "The Rise of the Nuclear Family in Early Modern England," in Charles E. Rosenberg, ed., *The Family in History* (Philadelphia: University of Pennsylvania Press, 1975), pp. 36–38, 56; David Hunt, *Parents and Children in History* (New York: Basic Books, 1970), pp. 85–86; and M. J. Tucker, "The Child as Beginning and End: Fifteenth and Sixteenth Century English Childhood," in Lloyd deMause, ed., *The History of Childhood* (New York: The Psychohistory Press, 1974), p. 238.

19. This point is very important, and his failure to understand it has made possible Lloyd deMause's attack on Ariès' work in his essay, "The Evolution of Childhood," pp. 1–73 of *The History of Childhood*. deMause calls Ariès' description of playing with infant genitals an example of sexual molestation, which such action certainly is when it occurs in the West today. But Ariès' whole point was that yesterday is *not* today; that sexual attitudes were very different then, and that the context of attitudes determines the meaning of an act. As for action that is unequivocally abusive, the point here is that love and hate have close ties. It is absence of contact that is the real psychological danger, for the child experiences such absence as apathy, and his psyche translates it as meaninglessness. Perhaps existential man's search for meaning originates in this tragic experience.

Second, the absence of ego in history never precluded violence, as the *Iliad* clearly shows. But such violence was of a very different order, it seems to me. It was moved by spontaneous passion; the concept of discipline as an institutional practice did not exist in schools prior to the sixteenth century (at least not corporal punishment), as Ariès notes. Such discipline is premediated, done for different reasons than immediate feelings. It is usually a form of sublimation, for example, sadism posing as self-righteousnes ("this hurts me more than it does you"). With the crystallization of an ego, emotions get twisted or transmuted into other forms. The result was already present in the monastic flagellant orders of the Middle Ages, and it was Reich's contention that much contemporary sexuality had a sadistic or masochistic edge to it, and vice versa. A brilliant elaboration of this theme was provided by Lindsay Anderson in his film *If...*, released in the late 1960s.

20. Montagu, *Touching,* p. 207. The discussion that follows is taken from pp. 60, 77–78, 120–24.

21. What type of men were Watson and Holt? According to a description provided by one of his colleagues and one of his assistants, Holt was the stereotype of Reich's armored individual. "His manner", they wrote, "was more than serious, it was earnest. There was nothing about him which could be called impressive, due perhaps to the absence of any outstanding feature; rather he appeared a highly efficient, perfectly coordinated human machine. He seemed to us austere and unapproachable." (Quoted in Montagu, *Touching,* p. 121). As for Watson, it is instructive to learn that not long after the publication of his *Psychology from the Standpoint of the Behaviorist* (1919) he accepted a job with the J. Walter Thompson advertising agency in New York, where "he applied his principles for controlling rats to the manipulation of consumers" (Philip J. Pauly, "Psychology at Hopkins," *Johns Hopkins Magazine* 30 [December 1979], 40).

22. Although I have tended to talk in causal terms in this discussion, I do not believe that the impact of child-rearing practices on adult life and culture is particularly more significant than the reverse. As indicated, I believe that the two form a historical gestalt, but the implications that follow are not fully clear. As Milton Singer shows in his "Survey of Culture and Personality," in Bert Kaplan, ed., *Studying Personality Cross-Culturally* (New York: Harper & Row, 1961), pp. 9–90, anthropology has had a difficult time trying to extricate itself from causal arguments while continuing to say something meaningful. Thus both Montagu and de-Mause talk as though this or that child-rearing practice results in this or that adult characteristic, but proof remains elusive and, in any event, theirs is a mechanical approach to very complex problems.

Some progress has been made by Gregory Bateson (see Chapter 7), whose analyses have tended to show that different kinds of interpersonal relations can assume functional patterns that differ from culture to culture. In this view, parent-child relations are part of the culturally patterned themes, and thus the child's relationship to its parents is mutually interactive, or holistic. Children are thus seen as active in stimulating parents into a certain pattern, a thesis supported by several of the studies in the volume by Evelyn Thoman cited in note 14. Margot Witty and T. B. Brazelton

made a similar argument in "The Child's Mind," *Harper's*, April 1978, pp.
46–47. The structure is seen to operate like a circuit rather than a line.

23. Marshall H. Klaus and John H. Kennell, *Maternal-Infant Bonding* (St. Louis: The C. V. Mosby Company, 1976), esp. pp. 58ff., and Louis W. Sander et al., "Change in Infant and Caregiver Variables over the First Two Months of Life: Integration of Action in Early Development," in Evelyn Thoman, ed., *Origins of the Infant's Social Responsiveness*, pp. 368–75. Popular coverage of the Klaus-Kennell work was provided by the *New York Times*, 16 August 1977, p. 30, under the title "Closeness in the First Minutes of Life May Have a Lasting Effect." Cf. Aidan Macfarlane, *The Psychology of Childbirth* (Cambridge: Harvard University Press, 1977), pp. 52–54, 100–101.

24. Montagu, *Touching*, pp. 256–58.

25. See Richard Poirier's interesting analysis of the lyrics, "Learning from the Beatles," in his book *The Performing Self* (New York: Oxford University Press, 1971), pp. 112–40.

26. From *The Child*, by Erich Neumann, p. 33. English translation copyright 1973 by the C. G. Jung Foundation for Analytical Psychology, Inc.

27. N. O. Brown, *Life Against Death* (Middletown, Conn.: Wesleyan University Press, 1970; first publ. 1959), p. 31.

28. There "may be another, less systematized, kind of memory [than the cognitive kind]," writes the pediatrician John Davies, "and it does not mean that the [preconscious] experience has been lost, or is not having an influence." Quoted in Macfarlane, *The Psychology of Childbirth*, p. 31.

29. C. G. Jung, "In Memory of Sigmund Freud," in *The Spirit in Man, Art, and Literature*, trans. R. F. C. Hull (Princeton: Princeton University Press, 1971), p. 48. This was Jung's obituary of Freud, originally published in 1939.

30. For Bowlby see his book *Separation* (New York: Basic Books, 1973). The quotation from Reich is on page 30 of *The Function of the Orgasm*, trans. Vincent R. Carfagno (New York: Pocket Books, 1975; orig. German edition 1942). There is some confusion here, as this book is vol. 1 of his *Discovery of the Orgone*, which he rewrote several times under the same title. In the discussion of Reich that follows, I have drawn on pp. 4–6, 15, 37, 88–96, 128–32, 162–69, 243–44, 269–71, 283 of this work, as well as from his book *Character Analysis*, trans. Vincent R. Carfagno, 3d ed., enl. (New York: Simon and Schuster, 1972; orig. publ. 1945), pp. 171–89.

31. I am not certain who coined this term, but it first becomes important in the anthropological literature in *The People of Alor*, by Cora DuBois, published in 1944. See Milton Singer, "A Survey of Culture and Personality," p. 33.

32. Actually, Reich said that it existed in all patriarchies, a thesis much

more difficult to establish. On Fromm's study of anal typology see "Die psychanalytische Charakterologie und ihre Bedeutung für die Sozialpsychologie," *Zeitschrift für Sozialforschung* 1 (1932), 253–77. The quote from Reich is in *Character Analysis,* p. xxvi.

33. Peter Koestenbaum, *Existential Sexuality* (Englewood Cliffs, N.J.: Prentice-Hall, 1974), pp. 63, 75.

34. Yankelovich and Barrett, *Ego and Instinct,* esp. pp. 157, 360–61, 365, 367–68, 371, 396. There is of course an immense specialized literature on cell development. A recent, somewhat popular article on cell interconnection is L. A. Staehelin and B. E. Hull, "Junctions between Living Cells," *Scientific American* 238 (May 1978), 141–52.

35. Itzhak Bentov, *Stalking the Wild Pendulum* (New York: Dutton, 1977), pp. 85–86. Another way of seeing this is to regard the brain as an organ like any other, whose function it is to amplify thoughts. What we call mind, which is identical to the body, goes from the top of the head to the bottom of the feet. The sensation of the body as an object of a consciousness localized in the head is a Cartesian illusion. Mircea Eliade has noted that premodern societies typically locate consciousness at a point just below the navel, which is also a classical yogic training exercise. Scientifically, this is probably more accurate than regarding consciousness as located inside the head. Naturally enough, modern culture regards it as located there because in a context so dominated by processing and control, the experience of rationality becomes overwhelming. In other contexts, it is less impressive. We become easily convinced that this mental processing is the most important, or even the only, form of thought. Bentov argues that it is not even *thought,* a position more or less taken by don Juan in his discussion with Carlos Castaneda in *Tales of Power* (see above, Chapter 5, note 19). Don Juan also maintains, as I have in the text, that *both* "tonal" and "nagual" are inherent parts of our being, but that the decisions we make occur in the realm of the nagual. However, he adds (p. 265) that the view of the tonal must prevail if one is to make use of the nagual—a point I regard as crucial in this whole business.

36. Aptly termed "split-brain follies" by Theodore Roszak. See *Unfinished Animal* (New York: Harper & Row, 1975), pp. 52–57. The various experiments that have been performed with the "two brains" might conceivably be seen as constituting a refutation of my argument about the body and unconscious knowledge. After all, these experiments do reveal the right hemisphere as being (in right-handed people) the locus of nonverbal functions. However, my argument does not deny that the brain stores images or organizes them. The "two brain" experiments tell us nothing about where the knowledge *originates* from. Thus I believe it can be maintained that intelligence is in the body, and data processing in the brain. Nor is this to deny that the brain can be a very sensual thing, amplifying and processing fantasy, dreams, artistic imagery, and so on.

37. Peter Marris, *Loss and Change* (Garden City, N.Y.: Doubleday Anchor, 1975).

38. There are limits to this argument, of course, but it is nevertheless likely that beyond the common substrate of primary process, Galileo's body was different from that of Thomas Aquinas, and that both of them were significantly different from the body of Homer. The human body has changed over the centuries in a number of important ways: in height, shape, ability to perceive colors, and especially in physiognomy. Psychoanalyst Stanley Keleman has developed this theme in some detail, and has argued that the body of the future will be a further radical departure from that of the present.

39. It should be clear that I have *not* left Descartes behind, largely because it is presently impossible to think discursively in purely nonscientific categories, although I have struggled to do so (cf. notes 35 and 36 to this chapter). The discussion in the text continues the mind/body dichotomy, locating the ego in the head and the unconscious in the body. It also uses the term "unconscious" in two senses, as participation, and as knowledge in the body which we somehow cannot get at. Can such an approach be justified?

I would answer by saying that this chapter has an inevitable tension built into it. I am trying to provide a verbal analysis of nonverbal experience, and there are obvious limits to what can be communicated in this way. As don Juan noted, the "tonal," by definition, cannot possibly explicate the "nagual." Thus the two senses of the unconscious which I make use of are only dual to scientific reasoning. To holistic reasoning, *mimesis is the knowledge present in the body*, and is hardly inaccessible. In other words, the "nagual" is not unknown. It is only unknown to the ego. The ontological being, the whole person, does know it, but there is no way of presenting this knowledge to the reader in book form short of having the text printed on fur or switching to verse. I could, of course, have invented a new holistic terminology, complete with words such as "mindbody" and "selfother," but I do not think a scientific *Finnegan's Wake* would be helpful at this point. I suggest, then, that the present chapter and its Cartesian vocabulary be viewed as a prop helping us to advance to the point where we shall no longer think in dualistic terms. We are still stuck in dualism, yet can recognize an approaching change.

40. E. A. Burtt, *The Metaphysical Foundations of Modern Science*, 2d ed. (Garden City, N.Y.: Doubleday, 1932), p. 17; Langer, *Philosophy in a New Key*, pp. xiii, 3, 12–13.

41. The following discussion is adapted from my essay, "The Ambiguity of Color," published in 1978 by the Exploratorium in San Francisco; use of this material by permission of the Director. See also Mike and Nancy Samuels, *Seeing with the Mind's Eye* (New York: Random House, 1975), p. 93. Land's article, "Experiments in Color Vision," may be found

in the May 1959 issue of *Scientific American,* and the quote from Lao-tzu appears in Alan Watts, *The Way of Zen* (New York: Vintage Books, 1957), p. 27. Whorf's classic work is *Language, Thought, and Reality,* ed. John B. Carroll (Cambridge: The MIT Press, 1956). There is a large literature on the human aura; the interested reader might start with Nicholas M. Regush, *Exploring the Human Aura* (Englewood Cliffs, N.J.: Prentice-Hall, 1975).

42. Exposure *beyond* fifteen minutes starts to push the inmate toward a breakdown. See "No new tortures needed," *Montreal Gazette,* 17 October 1980, and "Pink power calms raging inmates," *Montreal Gazette,* 5 January 1981.

43. This statement may be a bit misleading; I do not mean to suggest that anthropocentrism is the answer to our epistemological dilemmas. It is worth asking, for example, what the cetacean or arachnid experience of light and color is, and Judith and Herbert Kohl explore this approach in their interesting book, *The View from the Oak* (San Francisco: Sierra Club Books, 1977). Even in these cases, however, the human factor intrudes; what one is really studying is the human experience of the cetacean (or arachnid) experience of light and color. But recognizing the existence of this factor and incorporating it into our sciences does not necessarily result in anthropocentrism. Donald Griffin discusses the notion of participant observation in biological research in *The Question of Animal Awareness* (New York: The Rockefeller University Press, 1976).

44. Robert Bly, *Sleepers Joining Hands* (New York: Harper & Row, 1973), pp. 48–49.

45. Brown, *Life Against Death,* p. 236.

CHAPTER 7. *Tomorrow's Metaphysics (1)*

1. Philip Slater, *Earthwalk* (New York: Bantam Books, 1975), p. 233.

2. Gregory Bateson died in San Francisco in July 1980. He was working on a successor to *Mind and Nature,* which may have explored the aesthetic dimension that I discuss briefly in Chapter 9; but as it stands now, the discussion of his work in Chapters 7 and 8 below turns out, very unexpectedly, to be "complete."

A biography of Bateson appeared too late for me to read it for this work: David Lipset, *Gregory Bateson: The Legacy of a Scientist* (Englewood Cliffs, N.J.: Prentice-Hall, 1980).

3. The discussion of William Bateson's life given below is based on the following sources: William Coleman, "Bateson and Chromosomes: Conservative Thought in Science," *Centaurus* 15 (1970), 228–314; Beatrice Bateson's memoir of her husband, *William Bateson, F.R.S., Naturalist* (Cambridge: Cambridge University Press, 1928), pp. 1–160; and Gregory Bateson, *Steps to an Ecology of Mind* (London: Paladin, 1973; New York: Ballantine, 1972), pp. 47–52 British edition, 73–78 American edition.

4. Morris Berman, " 'Hegemony' and the Amateur Tradition in British Science," *Journal of Social History* 8 (Winter, 1975), 30–50. All British science, however, was colored by this tradition down to the late nineteenth century.

5. The full title is *Materials for the Study of Variation treated with especial Regard to Discontinuity in the Origin of Species.*

6. What was lost to science when chromosome theory triumphed we can only guess. Bateson's idea of the transmission of tendency has been revived in the work of Gregory Bateson, C. H. Waddington, and a few other biologists who have been able to argue successfully for the existence of Lamarckian mimicry—something that simulates the inheritance of acquired characteristics. But by and large, the world of materialistic, orthodox biology is leading, ineluctably, to the potential horrors of gene manipulation and recombinant DNA—horrors that might have been avoided had Bateson's views prevailed in the 1920s. Cf. Barry Commoner, "Failure of the Watson-Crick Theory as a Chemical Explanation of Inheritance," *Nature* 226 (1968), 334.

7. Victorian model-building, including the vortex atom, has been the subject of a large literature, including a very critical overview by the French historian Pierre Duhem in chapter 4 of his *Aim and Structure of Physcial Theory*, trans. Philip P. Wiener (Princeton: Princeton University Press, 1954; orig. French edition 1914). Further material can be obtained in works by and about William Thomson (Lord Kelvin), P. G. Tait, James Clerk Maxwell, Oliver Lodge, Joseph Larmor et al. Cf. Robert Silliman, "William Thomson: Smoke Rings and Nineteenth-Century Atomism," *Isis* 54 (1963), 461–74.

8. W. and G. Bateson, "On certain aberrations of the red-legged partridges *Alectoris rufa* and *saxatilis*," *Journal of Genetics* 16 (1926), 101–23.

9. Cf. Gunther S. Stent, *The Coming of the Golden Age* (Garden City, N.Y.: The Natural History Press, 1968), pp. 73–74, 112. See also his *Paradoxes of Progress* (San Francisco: W. H. Freeman, 1978).

10. By and large, I am going to omit any discussion of Bateson's biological writings and his revision of Darwinian evolution. Although integrally related to his other work, limitations of space prevent an exposition at this point. I am, also, primarily interested in the ethical implications of that work, and this is presented in Chapter 8. Readers interested in filling this gap should consult *Mind and Nature: A Necessary Unity* (New York: Dutton, 1979), and the essays in *Steps to an Ecology of Mind* titled "Minimal Requirements for a Theory of Schizophrenia" and "The Role of Somatic Change in Evolution."

11. The following discussion is taken from *Naven*, 2d ed. (Stanford: Stanford University Press, 1958), pp. 1–2, 29–30, 33, 35, 88, 92, 97–99, 106–34, 141–51, 157–58, 175–79, 186–203, 215, 218–20, 257–79, and the 1958 Epilogue. I have also used three articles from *Steps to an Ecology of Mind:*

Notes

"Experiments in Thinking about Observed Ethnological Material," "Morale and National Character," and "Bali: The Value System of a Steady State."

Bateson argues in *Mind and Nature*, pp. 192–95, that the methodology of the Iatmul investigation is a paradigm for the resolution of a very large number of problems in ethics, education, and evolution.

12. Bateson, however, had his differences with Ruth Benedict's approach, as he notes on pages 191–92 of *Mind and Nature*. The discussion that follows is concerned exclusively with ethos; I shall return to eidos in the section on learning theory, below.

13. There is, however, kinship differentiation, and naven turns out to be motivated by the attempt to reduce tensions (as personally experienced) arising from these relationships, in addition to its importance in resolving sexual tensions. For the most part, however, I shall not be dealing with kinship motivation. Bateson's summary can be found in *Naven*, pp. 203–17.

14. For an overview of some of the anthropological discussion on this topic, see Milton Singer, "A Survey of Culture and Personality," in Bert Kaplan, ed., *Studying Personality Cross-Culturally* (New York: Harper & Row, 1961), pp. 9–90.

15. It is necessary to note that Bateson's early anthropological work did contain two serious errors, both of which he later pointed out. The first was what Alfred North Whitehead called the "fallacy of misplaced concreteness"—the making of abstractions into concrete "things." Bateson was in fact aware of this when he wrote the Epilogue to the first (1936) edition of *Naven*. He states there that despite the way he tended to argue in the text, ethos is not an entity and cannot be the cause of anything: no one has ever seen or tasted an ethos any more than they have seen or tasted the First Law of Thermodynamics. The concept is a description, a way of organizing data, a viewpoint taken by the scientist or by the natives themselves.

Second, the notion that stability could be maintained by an "admixture" of symmetrical and complementary schismogenesis was, he realized by 1958, too rudimentary. It naïvely assumes the two variables can somehow cancel each other out, but never develops a functional relationship between them. Without such a relationship, there is no reason to expect that the two processes will equilibrate; the explanation for stability is much too fortuitous here. The real issue, Bateson saw later, was how (and whether) increasing schismogenic tension served to trigger controlling factors, and he came to reevaluate the theory in cybernetic terms with the concept of "end-linkage." Cf. Chapter 8 of the present work and the 1958 Epilogue to *Naven*.

16. On Bali, see Gregory Bateson and Margaret Mead, *Balinese Character: A Photographic Analysis* (New York: New York Academy of Sciences,

1942); the essay on Bali mentioned in note 11; and "Style, Grace and Information in Primitive Art," in *Steps to an Ecology of Mind.*

17. Herbert Marcuse, *One-Dimensional Man* (Boston: Beacon Press, 1964), p. 17.

18. The discussion below is based on Jurgen Ruesch and Gregory Bateson, *Communication: The Social Matrix of Psychiatry* (New York: Norton, 1968; orig. publ. 1951), pp. 176, 212, 218, 242; and the following articles from *Steps to an Ecology of Mind:* "Social Planning and the Concept of Deutero-Learning"; "A Theory of Play and Fantasy"; "Epidemiology of a Schizophrenia"; "Towards a Theory of Schizophrenia" (written together with Don D. Jackson, Jay Haley and John H. Weakland); "Minimal Requirements for a Theory of Schizophrenia"; "Double Bind, 1969"; and "The Logical Categories of Learning and Communication."

19. Bateson, *Steps to an Ecology of Mind,* p. 143 British edition; p. 170 American edition.

20. There is such a thing as a so-called lucid dream, in which the dreamer is aware that he or she is dreaming, but for the most part this phenomenon is not a common occurrence.

21. Jay Haley, "Paradoxes in Play, Fantasy, and Psychotherapy," *Psychiatric Research Reports* 2 (1955), 52–58.

22. R. D. Laing, *The Divided Self* (Harmondsworth: Penguin Books, 1965; first publ. 1959), pp. 29–30.

23. Quoted in Coleman, "Bateson and Chromosomes," p. 273.

24. See Bateson's Introduction to Gregory Bateson, ed., *Perceval's Narrative: A Patient's Account of His Psychosis, 1830–1832,* by John Perceval (Stanford: Stanford University Press, 1961).

25. E. Z. Friedenberg, *R. D. Laing* (New York: Viking, 1974), p. 7.

26. My source for the following information is a talk given by Bateson in London on 14 October 1975, and also pp. 121–23 of *Mind and Nature.*

27. *Steps to an Ecology of Mind,* p. 265 British edition, p. 295 American edition. For a delightful Victorian story based on this theme, see Edwin A. Abbott, *Flatland,* 6th ed. (New York: Dover, 1952).

28. R. D. Laing, *The Politics of Experience* (New York: Ballantine Books, 1968), pp. 144–45.

29. "Officially" is a key word here, since it is through metacommunication itself that we absorb the Cartesian world view. Cf. my discussion in Chapter 5, that the Cartesian metaphysics contains participating consciousness even while denying its existence.

CHAPTER 8. *Tomorrow's Metaphysics (2)*

1. Gregory Bateson, *Steps to an Ecology of Mind* (London: Paladin, 1973; New York: Ballantine, 1972), p. 31 British edition, p. xxv American edition.

2. I am using "Mind" here roughly in the sense first employed in Chapter 5, that is, to denote the mental system that includes both the unconscious and the mind (small m), or conscious awareness. The concept will be more fully elaborated in the discussion below.

3. For an interesting comparison with the following, see Jurgen Ruesch and Gregory Bateson, *Communication: The Social Matrix of Psychiatry* (New York: Norton, 1968; orig. publ. 1951), pp. 259-61.

4. On the following discussion, see "The Cybernetics of 'Self': A Theory of Alcoholism," in *Steps to an Ecology of Mind*.

5. It is interesting to note here that one of the founders of AA was influenced by the work of Carl Jung. See *Alcoholics Anonymous*, 3d ed. (New York: Alcoholics Anonymous World Services, 1976), pp. 26-27.

6. The following section is based on Bateson's *Mind and Nature: A Necessary Unity* (New York: Dutton, 1979), pp. 91-114, and *Steps to an Ecology of Mind*, p. 458 British edition, p. 482 American edition.

7. The discussion of redundancy given below is based on *Steps to an Ecology of Mind*, pp. 101-13 British edition, pp. 128-40 American edition.

8. Michael Polanyi, *Personal Knowledge*, corrected ed. (Chicago: University of Chicago Press, 1962), p. 88.

I hope I am not belaboring a point here, but it may not be immediately obvious that making everything redundant is equivalent to making everything random. A useful analogy might be the signal-to-noise ratio of a radio broadcast or TV screen: it must be a *ratio* if it is to exist at all. If everything were a signal, there would be no more background; so *everything* would be background (the TV screen would be black, for example). If every soldier in the army were promoted to the rank of general, there would be no more army. Total redundancy, in other words, destroys differentiation. When everything is redundant there is no longer a framework left to create redundancy. "If everybody is somebody," wrote Gilbert and Sullivan in one of their operettas, "then nobody is anybody."

9. Gregory Bateson, *Naven*, 2d ed. (Stanford: Stanford University Press, 1958), p. 276.

10. Gregory Bateson and Margaret Mead, *Balinese Character: A Photographic Analysis* (New York: New York Academy of Sciences, 1942). For representative kinesic studies, see R. L. Birdwhistell, *Introduction to Kinesics* (Louisville, Ky.: University of Louisville Press, 1952), and A. E. Scheflen, *How Behavior Means* (Garden City, N.Y.: Doubleday Anchor, 1974).

11. Anthony Wilden, *System and Structure* (London: Tavistock Publications, 1972), pp. 123, 194, and passim. On the discussion of analogue versus digital knowledge given below see *Steps to an Ecology of Mind*, pp. 109-12, 387-89, 408 British edition, and pp. 136-39, 411-14, 432-33 American edition.

12. Actually, I have some difficulties with Bateson's contention that the essence of an unconscious message is that it is unconscious, or that all analogue communication is an exercise in communication about the unconscious mind. Dance can be about the relationship between space and content, or lightness and gravity, for example. In the famous film *Les enfants du paradis* (Children of Paradise), Jean-Louis Barrault does a mime sequence about a pickpocket. The purpose of this sequence was not to reveal the nature of the unconscious, but to expose the theft of a watch. It is really straight psychodrama that Bateson is talking about, I think, rather than every type of analogue communication.

13. "A Conversation with Gregory Bateson," in Lee Thayer, ed., *Communication: Ethical and Moral Issues* (London and New York: Gordon and Breach Science Publishers, 1973), p. 248.

14. The discussion that follows is taken from "A Conversation with Gregory Bateson," p. 247; Mary Catherine Bateson, ed., *Our Own Metaphor* (New York: Knopf, 1972), pp. 16–17; John Brockman, ed., *About Bateson* (New York: Dutton, 1977), p. 98; *Psychology Today*, May 1972, p. 80 (interview with Lévi-Strauss); and *Steps to an Ecology of Mind*, pp. 95, 303, 410, 434–35, 459–60 British edition, and pp. 122–23, 332–33, 434, 460, 483–84 American edition. Cf. also Lynn White, Jr., "The Historical Roots of Our Ecologic Crisis," *Science* 155 (10 March 1967), 1203–7.

15. On acclimation versus addiction see *Mind and Nature*, pp. 172–74, 178, and *Steps to an Ecology of Mind*, pp. 321, 416–17, 465–5 British edition, and pp. 351, 441–42, 488–90 American edition.

16. Some of this information is available in the 1979 article by Pacific News Service (San Francisco), entitled "Global Comeback of Once-banished Malaria," by Rasa Gustaitis.

17. Sources for the following include the interview with Lévi-Strauss cited in note 14 of this chapter; Mary Catherine Bateson, *Our Own Metaphor*, pp. 91, 266–79, 285; *Steps to an Ecology of Mind*, pp. 420, 426, 475 British edition, and pp. 445, 451, 499 American edition; and Murray Bookchin "Ecology and Revolutionary Thought," in *Post-Scarcity Anarchism* (Palo Alto: Ramparts Press, 1971), esp. pp. 63–68, 70–82. The importance of diversity is also discussed in most textbooks of ecology or genetics.

CHAPTER 9. *The Politics of Consciousness*

1. Max Weber, *The Protestant Ethic and the Spirit of Capitalism*, trans. Talcott Parsons (New York: Scribner's, 1958; orig. German publ. 1904–5), p. 182.

2. *See* Chapter 7, note 24.

3. This statement may be wrong. The perception of colors in the human aura, and their relation to healing, may prove to be such an avenue of inquiry. Cf. my discussion of color at the conclusion of Chapter 6.

4. *Steps to an Ecology of Mind* (London: Paladin, 1973; New York: Ballantine, 1972), p. 436 British edition, p. 461 American edition.

5. Christopher Hill, *The World Turned Upside Down* (New York: Viking, 1972).

6. I am not personally active in "planetary politics" and therefore cannot speak with authority on these matters. What follows, then, should be understood as a report on certain trends that fall into this category. In this discussion, I shall be drawing on the literature cited below to construct the argument, but I wish to state my great indebtedness to Peter Berg for opening my eyes to the subject in general. Many of the ideas presented in the discussion that follows have been the focus of his own political and educational efforts in the San Francisco Bay Area for more than a decade now, through his journal *Planet Drum*, his book *Reinhabiting a Separate Country* (San Francisco: Planet Drum Books, 1978), and numerous other activities. The nature of planetary culture, and its existence as a political alternative, was also the subject of a four-day conference titled "Listening to the Earth," which was codirected by Berg and myself in San Francisco April 7–10, 1979. Some of the discussion presented below draws on ideas articulated at this conference.

The general literature on the subject is fairly large at this point, so I can do no more than cite my favorites:

Fiction: Ernest Callenbach, *Ecotopia* (Berkeley: Banyan Tree Books, 1975); Ursula K. LeGuin, *The Dispossessed* (New York: Avon, 1974); Marge Piercy, *Woman on the Edge of Time* (New York: Knopf, 1976).

Futures research: Willis W. Harman, *An Incomplete Guide to the Future* (San Francisco: San Francisco Book Company, 1976); Kimon Valaskakis et al., eds., *The Conserver Society* (New York: Harper & Row, 1979); Hazel Henderson, *Creating Alternative Futures* (New York: Berkley Publishing Corp., 1978); Edward Goldsmith et al., eds., *Blueprint for Survival* (Boston: Houghton Mifflin, 1972); Peter Hall, ed., *Europe 2000* (London: Gerald Duckworth, 1977); Michael Marien, "The Two Visions of Post-Industrial Society," *Futures* 9 (1977), 415–31, and "Toward a Devolution," *Social Policy*, Nov./Dec. 1978, pp. 26–35.

Political and economic commentary: Leopold Kohr, *The Breakdown of Nations* (New York: Dutton, 1975; orig. publ. 1957); Gary Snyder, "Four Changes," in *Turtle Island* (New York: New Directions, 1974); Gordon Rattray Taylor, *Rethink* (Harmondsworth: Penguin Books, 1972); Michael Zwerin, *Case for the Balkanization of Practically Everyone* (London: Wildwood House, 1975); E. F. Schumacher, *Small is Beautiful* (New York: Harper & Row, 1973); Herman E. Daly, ed., *Toward a Steady-State Economy* (San Francisco, W. H. Freeman, 1973); "Ecology Party Manifesto," *The New Ecologist* 9 (1979), 59–61; and the literature of a number of anarchist writers and/or social critics, particularly Paul Goodman, Ivan Illich, Lewis Mumford, and Murray Bookchin (on the connection between anarchism and ecology, see the essay by Bookchin entitled "Ecology and Revolutionary Thought," in *Post-Scarcity Anarchism* [Palo Alto: Ramparts Press, 1971], and also George Woodcock, "Anarchism and Ecology," in *The Ecologist* 4 [1974], 84–88).

Ecology: Arne Naess, "The Shallow and the Deep, Long-Range Ecology Movement. A Summary," *Inquiry* 16 (1973), 95–100; Paul Shepard, *The Tender Carnivore and the Sacred Game* (New York: Scribner's, 1973); Bill Devall, "Streams of Environmentalism," *Natural Resources Journal* 19 (1979), no. 3; John Rodman, "The Liberation of Nature?" *Inquiry* 20 (1977), 83–131; Raymond F. Dasmann, "Toward a Dynamic Balance of Man and Nature," *The Ecologist* 6 (1976), 2–5, and "National Parks, Nature Conservation and 'Future Primitive,'" *The Ecologist* 6 (1976), 164–67; and the marvelous essay by Jerry Gorsline and Linn House, "Future Primitive," which appeared in *Planet Drum*, no. 3 ("Northern Pacific Rim Alive"), 1974, and was reprinted in *Alcheringa* 2 (1977), 111–13.

Religious renewal: Eleanor Wilner, *Gathering the Winds* (Baltimore: The Johns Hopkins University Press, 1975), and Jacob Needleman, *A Sense of the Cosmos* (Garden City, N.Y.: Doubleday, 1975).

7. Unfortunately, parapsychology has, for many years now, been taken quite seriously by the CIA and the KGB, which are interested in its possible military applications. I discuss the political dangers of Learning III later in this chapter, but the reader should be aware of the heavy American and Soviet investment in psychic research per se, most of it classified information. There have been a few public revelations of CIA experimentation with LSD (for example, Project MK-ULTRA) as a result of material being released under the recent Freedom of Information Act, but otherwise little is available. See Michael Rossman, *New Age Blues* (New York: Dutton, 1979), pp. 167–260; John D. Marks, *The Search for the "Manchurian Candidate": The CIA and Mind Control* (New York: Times Books, 1979); "Soviet Psychic Secrets," *San Francisco Chronicle*, 16 June 1977; John L. Wilhelm, "Psychic Spying?" *Washington Post*, 7 August 1977.

8. Fernand Lamaze, *Painless Childbirth* (New York: Pocket Books, 1977); Frederick Leboyer, *Birth Without Violence* (New York: Knopf, 1975).

9. I am using "self" here in the Jungian sense rather than in Bateson's sense of ego (see Chapter 8).

10. The American video artist Paul Ryan has been working on just such an experiment for years, which he calls "triadic practice," in which groups of threes learn to avoid escalation to conflict. Some aspects of his work are dealt with in his book *Cybernetics of the Sacred* (Garden City, N.Y.: Doubleday Anchor, 1974), and more explicitly in "Relationships," *Talking Wood* 1 (1980), 44–55.

11. Jerry Gorsline and Linn House, "Future Primitive."

12. Murray Bookchin, *Post-Scarcity Anarchism*, p. 78; Gary Snyder, "Four Changes," p. 94.

13. There is by now a large literature on what is called "appropriate technology," or "technology with a human face." Two of the most well-known works are Ivan Illich, *Tools for Conviviality* (New York: Harper & Row, 1973), and Schumacher, *Small is Beautiful.*

14. Murray Bookchin, *The Limits of the City* (New York: Harper & Row, 1973); Lewis Mumford, *The Culture of Cities* (New York: Harcourt, Brace and Company, 1938). The quote from Ariès is in *Centuries of Childhood,* trans. Robert Baldick (New York: Vintage Books, 1962), p. 414.

15. Peter Berg and Raymond Dasmann, "Reinhabiting California," in Peter Berg, *Reinhabiting a Separate Country,* p. 219.

16. Roszak's work, especially *Unfinished Animal* (New York: Harper & Row, 1975), *Where the Wasteland Ends* (Garden City, N.Y.: Doubleday Anchor, 1972), and *Person/Planet* (Garden City, N.Y.: Doubleday, 1978), is premised on the Roman Empire model. Cf. Harman, *Incomplete Guide,* and Robert L. Heilbroner, *Business Civilization in Decline,* (New York: Norton, 1976).

17. Percival Goodman, *The Double E* (Garden City, N.Y.: Doubleday Anchor, 1977).

18. "A Future That Means Trouble," *San Francisco Chronicle,* 22 December 1975.

19. Leopold Kohr, *The Breakdown of Nations;* Kevin Phillips, "The Balkanization of America," *Harper's,* May 1978, pp. 37–47; Peter Hall, *Europe 2000,* esp. pp. 22–27 (in general, all of the trends I have sketched in the vision of a planetary culture are laid out in this book, including some of the sources for change). See also Zwerin, *Case for the Balkanization of Practically Everyone.*

20. Hall, *Europe 2000,* p. 167.

21. *I Ching,* trans. Richard Wilhelm and Cary F. Baynes, 3d ed. (Princeton: Princeton University Press, 1967), p. 186 (Hexagram 48, The Well).

22. William Coleman, "Bateson and Chromosomes: Conservative Thought in Science," *Centaurus* 15 (1970), esp. pp. 292–304.

23. Bateson's work has important political implications, but these are not particularly emphasized or utilized in his own analyses. In the case of schizophrenia, for example, the largest unit of Mind under his consideration is the family, and the family is hardly isolated from a wider political context. Authority relationships of the larger society are duplicated within the family structure, but this problem is never addressed. The power relationship that obtains between parent and child does not always produce schizophrenia, of course, but as Bateson pointed out in 1969, it is a necessary condition for it: the victim must be unable to leave the field. Hence Bateson's focus on disturbances of metacommunication is important in the analysis, but perhaps incomplete.

24. Anthony Wilden, *System and Structure* (London: Tavistock Publications, 1972), p. 113.

25. Anatol Rapoport, "Man, the Symbol User," in Lee Thayer, ed., *Communication: Ethical and Moral Issues* (New York: Gordon and Breach Science Publishers, 1973), p. 41.

26. The following critique of logical typing and hierarchy is the contribution of Paul Ryan, and I am grateful for his help in this difficult area. I hasten to add that Ryan does not share the other criticisms of Bateson's work which I have presented in this chapter.

For an elaboration of Ryan's critique, see "Metalogue: Gregory Bateson/Paul Ryan," in a special issue (Spring 1980) of the magazine *Talking Wood* titled "All Area."

27. G. Spencer Brown, *The Laws of Form* (New York: The Julian Press, 1972), p. x.

28. Warren S. McCulloch, "A Heterarchy of Values Determined by the Topology of Nervous Nets," in *Embodiments of Mind* (Cambridge: The MIT Press, 1965), pp. 40–44.

29. Hierarchy certainly exists in the animal kingdom, as shown by various studies of the behavior of wolf packs and other animal groups; but there is no way to prove the existence of a carry-over to human nature, as advocates of class society frequently wish to do. In Murray Bookchin's opinion, there "are no hierarchies in nature other than those imposed by hierarchical modes of human thought, but rather differences merely in function between and within living things" (*Post-Scarcity Anarchism,* p. 285). Some of Henri Laborit's work argues this view as well.

Strictly speaking, heterarchy and egalitarianism are not the same thing. Heterarchy is intransitive differentiation, which is not identical to equality. But the two are so close that in actual practice a heterarchical system would be virtually egalitarian.

30. René Dubos, "Environment," *Dictionary of the History of Ideas,* 2 (1973), 126; C. H. Waddington, "The Basic Ideas of Biology," in C. H. Waddington, ed., *Towards a Theoretical Biology,* 4 vols. (Chicago: Aldine Publishing Company, 1968), 1: 12.

31. Wilden, *System and Structure,* pp. 141, 354ff. Shannon, Weaver, and W. Ross Ashby are typical of the early cybernetic writers.

32. Of course, this is a tricky issue. Whether a change was a true alteration of a program, or part of the program all along, is a subject that historians debate for nearly every major historical development. The whole quantity-to-quality argument developed by Marx was designed to overcome the tension between homeorhetic and morphogenetic development.

33. Bateson, *Mind and Nature: A Necessary Unity* (New York: Dutton, 1979), p. 206.

34. Robert Lilienfeld, *The Rise of Systems Theory* (New York: Wiley, 1978), p. 70.

35. Ibid., p. 160.

36. Ibid., pp. 174, 263. Cf. William W. Everett, "Cybernetics and the Symbolic Body Model," *Zygon* 7 (June 1972), 104, 107.

37. Carolyn Merchant, *The Death of Nature* (New York: Harper & Row, 1980), pp. 103, 252, 291; see also pp. 238–39.

38. In point of fact, the experiment did not work as described. As Bateson tells us, the situation was so often on the verge of breaking down that the trainer had to give the animal numerous rewards to which it was not entitled in order to maintain his relationship with it.

39. See Chapter 7, note 27.

40. Rossman, *New Age Blues*, pp. 54–56. On the following paragraph cf. Chapter 7, note 2.

41. According to Flo Conway and Jim Siegelman, *Snapping: America's Epidemic of Sudden Personality Change* (Philadelphia: Lippincott, 1978), pp. 11–12, 56, 161, there are currently more than one thousand religious cults now active in the United States, using nearly eight thousand techniques that fall under the rubric of what Bateson calls Learning III. Many are run or guided by Madison Avenue experts, and the following in these cults is not necessarily small: the Church of Scientology, for example, has an estimated 3.5 million members in America alone.

42. Jerry Mander, *Four Arguments for the Elimination of Television* (New York: William Morrow, 1978), pp. 100–107.

43. Rossman, *New Age Blues*, p. 117.

44. Max Horkheimer, *Eclipse of Reason* (New York: The Seabury Press, 1974; orig. publ. 1947), p. 120. For material on *est* see Rossman, *New Age Blues*, pp. 115–66; Peter Marin, "The New Narcissism," *Harper's*, October 1975, pp. 45–56; Suzanne Gordon, "Let Them Eat *est*," *Mother Jones* 3 (December 1978), 41–54; and Jesse Kornbluth, "The Führer over est," *New Times* 6 (19 March 1976), 36–52.

45. On the link between Nazism and the occult, see Jean-Michel Angebert, *The Occult and the Third Reich*, trans. Lewis Sumberg (New York: Macmillan, 1974); Trevor Ravenscroft, *The Spear of Destiny* (New York: Putnam's 1973); and Dusty Sklar, *Gods and Beasts* (New York: Crowell, 1977).

46. Lucien Goldmann, *Immanuel Kant*, trans. Robert Black (London: New Left Books, 1971; orig. German publ. 1945, rev. ed. [French] 1967), p. 122; reprinted with permission of the publisher.

47. William Irwin Thompson, "Notes on an Emerging Planet," in Michael Katz et al., eds., *Earth's Answer* (New York: Harper & Row, 1977), p. 211.

48. Ibid., p. 213. I have fudged a bit here; Thompson is referring not to his own statement, but to that of Jonas Salk in his book, *The Survival of the Wisest*. Unfortunately, there is not much difference between the two. Thompson's own statement necessarily involves a distinction between shepherds and flocks, which he does not seem to see.

49. Julian Jaynes, *The Origin of Consciousness in the Breakdown of the Bicameral Mind* (Boston: Houghton Mifflin, 1976).

50. Bruce Brown, *Marx, Freud, and the Critique of Everyday Life* (New York: Monthly Review Press, 1973), p. 17.

51. Horkheimer, *Eclipse of Reason*, pp. 122–23.

52. See the two articles by Dasmann in *The Ecologist* for 1976, cited in note 6 to this chapter.

53. Gorsline and House, "Future Primitive." Berg defines a bioregion as "a geographical area united by particular natural characteristics (plants, animals, soils, watersheds, climate) and by human influences that bear on the region" (personal communication).

54. Marshall Sahlins, *Stone Age Economics* (Chicago: Aldine Publishing Company, 1972).

55. Robert Curry discusses the map in "Reinhabiting the Earth: Life Support and the Future Primitive," *Truck*, no. 18 (1978), pp. 17–40. The map is reproduced on page 190 of the same issue, and was originally part of Occasional Paper no. 9 of the International Union for Conservation of Nature and Natural Resources (Morges, Switzerland). Curry's article may also be found in John Carins, ed., *The Recovery of Damaged Ecosystems* (Blacksburg, Va.: Virginia Polytechnic University Press, 1976).

56. Berg and Dasmann, "Reinhabiting California," pp. 217–18.

57. Ibid., p. 217.

58. Mander, *Four Arguments*, pp. 104–5. Blake was making the same point when he wrote: "Earth and all you behold: tho' it appears without, it is within."

Although this distinction between nature-based religions and guruism is crucial, ecology is probably not by itself a sufficient guarantee against fascism, as Daniel Cohn-Bendit points out in a recent interview in *Le Sauvage* (No. 57, septembre 1978, p. 11). In June 1978, he notes, one Hamburg ecological party was openly fascist, taking the line of "Blood and Soil," and combining its antinuclear stance with a platform that was antigay, antifeminist, anti-Semitic, etc., and highly nationalistic. Although, as I have indicated, regionalism is intrinsically opposed to nationalism, in practice the line gets somewhat slippery. This was certainly the case in France, where regionalist proponents such as Charles Maurras wound up supporting the Vichy government.

59. Gorsline and House, "Future Primitive."

60. Wilden, *System and Structure*, pp. 21, 25; Jacques Lacan, "The Mirror Phase," trans. Jean Roussel, *New Left Review*, no. 51 (1968; orig. French version 1949), pp. 71–77.

61. Robert Bly, "I Came Out of the Mother Naked," in *Sleepers Joining Hands* (New York: Harper & Row, 1973), pp. 29–50.

62. Homer, *The Odyssey*, Book XI.

63. *Rebel in the Soul* was translated most recently by Bika Reed (New

York: Inner Traditions International, 1978), and excerpts are reprinted here
with the permission of the publisher. The first translation into a European
language was into German by A. Erman in 1896, and there have been a
number of others, for example, John A. Wilson's translation, "A Dispute
Over Suicide," in James B. Pritchard, ed., *Ancient Near Eastern Texts Relating to the Old Testament*, 3d ed. (Princeton: Princeton University Press,
1969; orig. publ. 1950), pp. 405–7, or Hans Goedicke, *The Report About the
Dispute of a Man with his Ba* (Baltimore: The Johns Hopkins University
Press, 1970).

Julian Jaynes discusses the document on pp. 193–94 of *Origins of Consciousness*, arguing that its language is not what the translators took it to
be, namely genuine self-dialogue. Thus he writes that "all translations of
this astounding text are full of modern mental impositions," whereas what
is really going on is auditory hallucination. Though it is true that there are
as many translations as there are translators, I believe Jaynes is somewhat
confused. He argues that the voice of the soul here cannot be a modern
one, in that bicameral consciousness mandates that we translate it as auditory hallucination; yet he also argues that the document dates from a
period of societal breakdown, and that it was in such periods that bicameral consciousness also broke down and ego consciousness emerged. But
this means precisely that gods, or auditory hallucinations, are converted
into selves, or interior voices. For this reason, I think we can take the
contemporary translations as accurate.

64. Paul Shepard, *The Tender Carnivore and the Sacred Game*, pp. 125, 283.

Glossary

Alembic: Egg-shaped glass container with a tube extending from the top. A standard piece of alchemical laboratory equipment in which many of the essential operations of alchemy, especially distillation, took place.

Analogue knowledge: Also called iconic communication. The range of non-verbal (excepting poetry), affective communication and perception by which we come to know the world, including fantasy, dreams, art, body language, gesture, and intonation. Contrasted with *Digital knowledge,* which is verbal-rational and abstract. Cf. Dialectical reason, Kinesics, Primary process.

Animism: Belief that everything, including what we commonly regard as inert material objects, is alive, possesses an indwelling spirit.

Archaic tradition: Used in this work interchangeably with the following terms: esoteric tradition, sympathy/antipathy theory, Hermetic tradition, Homeric or pre-Homeric mentality, *mimesis* (qv), animism (qv), totemism, participation, original participation, gnosticism, doctrine of signatures, and participating consciousness.

Strictly speaking, these terms are not identical. For example, the Hermetic tradition includes alchemy, which was probably not practiced during the pre-Homeric period, and which certainly postdates animism and totemism. Nor is all participating consciousness necessarily original (animistic).

However, common to all these terms is the notion that in a literal or figurative sense, everything in the universe is alive and interrelated, and that we know the world through direct identification with it, or immersion in its phenomena (subject/object merger). The archaic tradition, however, is not one of pure phenomenology, for it assumes the existence of natural laws or relationships that human beings can learn as a science. Among the most ancient of these sciences is totemism, the perceptible manifestation of indwelling spirits by icons or carved images. The medieval science of these correspondences—whereby plants, animals, minerals, parts of the body, and so on were seen as consciously

343

displaying the influence of particular stars or planets—was called the doctrine of signatures. Sympathetic magic was also based on the theory that certain things naturally went with (were sympathetic to) certain other things.

Atomism: The doctrine, which includes material atomism, that any phenomenon or object is no greater or less than the sum of its parts. It assumes a phenomenon is explained when it has been broken down into its constituent parts, which can then (at least theoretically) be reassembled. Contrasted with Holism (qv).

Cartesian paradigm: Dominant mode of consciousness in the West from the seventeenth century to the present. Defines as real that which can be analyzed or explained by the scientific method, a set of procedures combining experiment, quantification, atomism (qv), and the mechanical philosophy. The world is seen as a vast collection of matter and motion, obeying mathematical laws.

Circuitry: In cybernetic theory, the interrelation of parts, or of message exchange. The principle of circuitry holds that no variation can occur in one part of the system or circuit without setting off a chain reaction that is felt at every other point.

Coding: The programming or standarization of a person by his or her culture into its ethos (qv) and eidos; also, the program or mode of organization of the culture at large. See also Learning II, Tacit knowing, Gestalt, Paradigm. In cybernetic theory, coding refers to the translation of information into a set of symbols for meaningful communication.

Context: Stated or unstated set of rules within which an event or relationship takes place.

Cybernetics: Study of human control functions and the machines designed to replace them. More broadly, the science of messages, information exchange and communication.

Deutero-learning: see Learning II.

Developmentals: Incomplete psychic structures, such as ego and language, that are innate in the human being in embryonic or potential form. In order to be realized, their program of biological development must interact with particular social or cultural experiences at a specific stage in the life cycle.

Dialectical reason: Mode of analysis that sees things and their opposites as related. In this view, love and hate, or resistance and attachment, are not opposites but two sides of the same coin. The logic of dreams, or primary process (qv), is dialectical.

Digital knowledge: see Analogue knowledge.

Eidos: see Ethos.

Entropy: Measure of randomness, or disorganization. The opposite is negative entropy, or information. A system is said to have meaning when it gives us information, and it has such meaning when pattern, or redundancy, is present.

Epistemology: Branch of philosophy that attempts to determine the nature of knowing, or what the human mind can legitimately hope to discover about the objective world. The study of how the mind knows what it knows.

Ethos: Overall emotional tone of a culture; its affective paradigm (qv), or system of sentiments, as opposed to the *Eidos*, which is its cognitive paradigm, or intellectual world view. Eidos thus refers to the reality system of a culture, whereas ethos approximates the "etiquette," or norms of cultural behavior.

Fact-value distinction: Consciousness of the modern scientific era, according to which the good and the true are not necessarily related; value or meaning cannot be derived from data or empirical knowledge.

Feedback: In cybernetic theory, the use of part or all of the output of a system (e.g., a system of temperature control in a house) as input for another phase. Negative or self-corrective feedback, which is obtained by feeding the results of past actions back into a system, enables the system to maintain homeostasis (qv); such a situation is also called optimization. In a runaway situation, the feedback is positive, or escalating, building to a climax over time. In this situation the system is attempting to maximize certain variables rather than optimize them. Cf. Circuitry.

Figuration: Formation of mental pictures or images from the data of pure sensations. If I smell coffee and the picture of a cup of coffee suddenly comes to mind, I can be said to have figurated it.

Gestalt: A totality of interlocking imagery or concepts having specific properties that cannot be derived from its component parts. A pattern or world view that possesses a certain unity. Cf. Holism.

Holism: Also called synergy, or the synergistic principle. Holds that a collection of entities or objects can generate a larger reality not analyzable in terms of the components themselves; that the reality of any phenomenon is usually larger than the sum of its parts.

Homeostasis: Tendency of any system to maintain or preserve itself, to return to status quo if disturbed. A homeostatic system is steady state: it seeks to optimize rather than maximize the variables within it. Cf. Circuitry, Feedback.

Iconic communication: see Analogue knowledge.

Immanence: Doctrine that God is present within the phenomena we see, rather than external to them. Pantheism, animism (qv), and Batesonian holism are all variations on this theme. Contrasted with Transcendence, which sees God in heaven, external to the phenomena around us. Cartesianism and mainstream Judeo-Christian thinking fall into this category.

Individuation: According to Carl Jung, a process of personal growth and integration whereby a person evolves his true center, or Self, as opposed to his ego. The ego, or persona, is seen as the center of conscious life, whereas the Self is the result of bringing the conscious mind into harmony with the unconscious.

Kinesics: Study of body language and nonverbal communication, including posture, gesture, and movement, as clues to human personality and interaction.

Lapis-Christ parallel: Analogy between Christ and the work of alchemy. This was part of the claim, occasionally made in the Middle Ages, that alchemy was the inner content of Christianity, and that the manufacture of the philosopher's stone (*lapis*) was equivalent to the Christ-experience.

Learning I: The simple solution of a specific problem.

Learning II: Progressive change in the rate of Learning I. Understanding the nature of the context (qv) in which the problems posed in Learning I exist; learning the rules of the game. Equivalent to paradigm (qv) formation.

Learning III: An experience in which a person suddenly realizes the arbitrary nature of his or her own paradigm (qv), or Learning II, and goes through a profound reorganization of personality as a result. This change is usually experienced as a religious conversion, and has been called by many names: *satori*, God-realization, oceanic feeling, and so on.

Meristic differentiation: Repetition of like parts or segments along the axis of an animal, as in the earthworm.

Metacommunication: Communication about communication. "What is the nature of this conversation?" is a metacommunicative statement.

Metamerism: Dynamic asymmetry, or serial difference, between the successive segments of the parts of an animal; the claw of the lobster, for example. The animal displaying metamerism generally has most of its parts similar to each other, as in meristic differentiation, but with some marked by special asymmetric development.

Mimesis: Greek word for imitation, and the root of English words such as "mime" and "mimicry." More broadly, submitting to the spell of a performer, or becoming immersed in events; the state of consciousness in which the subject/object dichotomy breaks down and the person feels identified with what he or she is perceiving. Also called participating consciousness. It includes original participation, but is not necessarily animistic. See Archaic tradition.

Nonparticipating consciousness: State of mind in which the knower, or subject "in here," sees himself as radically disparate from the objects he confronts, which he sees as being "out there." In this view, the phenomena of the world remain the same whether or not we are present to observe them, and knowledge is acquired by recognizing a distance between ourselves and nature. Also called subject/object dichotomy.

Original participation: see Animism.

Paradigm: A world view or mode of perception; a model around which reality is organized. Cf. Gestalt.

Participation, or *Participating consciousness:* see Archaic tradition, *Mimesis.*

Prima materia, or *Materia prima:* Literally, first matter. In alchemy, it was the formless substance that resulted when a metal was dissolved, and from which the alchemical work of coagulation or recrystallization was begun. In allegory or personal growth (see Individuation), the stage of chaos from which a new form or personality will eventually congeal.

Primary process: Thought patterns associated with the unconscious, such as dream imagery, as opposed to rational ego-consciousness, or secondary process. See Archaic tradition.

Principle of incompleteness: Theory that most of our knowledge of the world is tacit in nature (see Tacit knowing) and thus that it has an ineffable basis, as a result of which it cannot be described in any rationally coherent sense. Furthermore, the principle sees the process of reality itself as ontologically incomplete. This theory is directly opposed to the Cartesian paradigm which holds that the mind can know all of reality; and also to the Freudian view, that all unconscious material can and should be made conscious.

Proto-learning: see Learning I.

Radical relativism: A possible consequence of the sociology of knowledge, that if all realities or methodologies are a product of specific historical circumstances, then all truth is relative to its individual context and there is no absolute or transcultural truth. This also implies that any given epistemology or world view is as accurate, or no less accurate, than any other.

Second Law of Thermodynamics: States that everything naturally tends toward entropy (qv). It is for this reason that the creation of information, or meaning, is seen as requiring effort.

Shadow: In Jungian terminology, the repressed and unconscious part of the personality which has to be recognized and integrated by the conscious mind in the process of individuation (qv). More broadly, the shadow is the undeveloped side of any natural pair of character traits. Men typically have a feminine shadow ("anima") and women a masculine one ("animus"); sadists also possess a streak of masochism; very serious persons have an unexpressed frivolous side, and so on.

Solve et coagula: Literally, dissolve and coagulate, a phrase summarizing the essence of the alchemical process. This involves reduction to the *prima materia* (qv) and then gradual fixation into a new pattern.

Steady state: Homeostasis (qv). The term is also used to refer to any type of nonprofit-oriented economy (e.g., feudalism) that does not expand over time, but only seeks to maintain itself.

Tacit knowing: Subliminal awareness and comprehension of information, especially information about the particular paradigm (qv) into which a given person is born. This operates on a gestalt and unconscious level, and consists of the ethos (qv) of a culture as well as the eidos. The concept of tacit knowing presupposes that any articulated world view is the result of unconscious factors that are culturally as well as biologically filtered and influenced. See also Gestalt, Principle of incompleteness, Figuration, Analogue knowledge.

Teleological: Pertaining to purpose, or goal. Aristotelian physics is termed teleological because it argues that objects fall to earth because they seek it as their natural place.

Teratology: Study of monstrosities, or abnormal formations in the animal and plant kingdoms.

Theory of Logical Types: As formulated by Alfred North Whitehead and Bertrand Russell, this theory states that no class of objects, as defined in logic or mathematics, can be a member of itself. As a logical construct, for example, we can form a class consisting of all the elephants that exist in the world. The theory states that this construct is not itself an elephant; it has no trunk, and eats no hay. The essential point of the theory is that there is a fundamental discontinuity between a class and its members.

Transcendence: see Immanence.

Trans-contextual: The characteristic of seeing things or situations as having a symbolic as well as a literal dimension. Madness, humor, art, and poetry are all trans-contextual in nature, operating on the level of metaphor or "double take."

Transform: In cybernetic theory, a change in the structure or composition of information without any corresponding alteration in meaning. Cf. Coding.

Index

349

Index

Boehme, Jacob, 84, 124
Brown, Norman O., 148, 172, 315

Cabala, 96, 100–101, 103
Capital accumulation, 49–50, 55, 57, 112–13, 123–24
Capitalism, 57; and ego development, 159–60
Cartesian logic, 32–36, 320, 329, 332–33, 344; and Freud, 172–73; and Polanyi, 155; and quantum mechanics, 144–45, 148; and schizophrenia, 35, 45–46, 271. *See also* Descartes, René
Cartesian physics, 110–11
Castaneda, Carlos, 94, 295, 311, 328
Character armor, 126–29, 173–76, 317, 326
Child development, 157–61, 165–70, 178–79, 324–26
Chromosome theory, 198–200, 331
Chromo-therapy, 186, 330
Circuitry, 200, 255–57, 259, 344. *See also* Cybernetics
Coding, 215, 344
Cognitive dissonance, 137
Color, 186–87, 329–30, 335; and alchemy, 85; and Edwin Land, 185–87; and Newton, 44–45
Copernicus, Nikolaus, 54
Croll, Oswald, 74
Cultism, 289–91, 340
Cybernetics, 200, 202, 239–49, 255–57, 259, 273–74, 282–87, 337, 344

Dali, Salvador, 96, 189; *The Persistence of Memory*, 98
Darwinian theory, 197, 202, 258–59, 331
Dasmann, Raymond, 293–95
De Occulta Philosophia. See Agrippa von Nettesheim
De Re Metallica. See Agricola, Georg
Dee, John, 103, 303
Della Porta, 74, 103
Delmedigo, Joseph Solomon, 100, 101, 103, 313
Descartes, René, 24, 28–29, 31–36, 82, 111, 303, 329; works of: *Discourse on Method*, 28, 31–32; *Meditations on First Philosophy*, 32–33, 36; *Principles of Philsophy*, 34, 111
Deutero-learning. *See* Learning II
Developmentals, 163, 178–81, 344
Dialectical reason, 344
Digital knowledge, 218–19, 254, 269, 344. *See also* Analogue knowledge

Discourse on Method. See Descartes, René
Diversity, ethics of, 255, 262–64, 335
Don Quixote (Cervantes), 75–76, 309
Dorn, Gerhard, 187
Double bind, 222–23, 224–26, 227–31, 272
Dreams, 309–10, 333; and Descartes, 36; and Jung, 78–84, 156; and Wilhelm Reich, 176–77
Drug use, 21, 171, 302, 337

Eckhart, Meister, 89
Ecology, 149, 276, 293–95, 341
Economy: Commercial Revolution, 50, 53–55; England, 16th century, 122; feudal, 52–53, 56, 58–59, 305–6; planetary, 277; Renaissance, 53–59. *See also* Capital accumulation
Ego: consciousness of, 162; crystallization of, 159, 162–65, 170, 323; development of, 158–65, 170, 179, 323; knowledge of, 146; psychology of, 323
Eidos, 205–6, 332, 344
Einsteinian physics, 143
Eliade, Mircea, 56, 88, 89, 328
Elim (J. S. Delmedigo), 101–3, 313
Eliot, T. S., 85
Eluard, Paul, 153
English Civil War, 95, 122
English Restoration, 123–24
Enthusiasm, 122–24; attack on, 129–31
Entropy. *See* Second Law of Thermodynamics
Epimenides' Paradox, 220–21, 321–22
Escher, M. C.: *Three Worlds*, 258
"est," 289–91, 295; Werner Erhard and, 289
Ethos, 205–6, 332, 345

Fact-value distinction, 40, 187–89, 193–94, 216–17, 233–34, 314
False-self system, 19–20, 35
Fascism: and the occult, 290–92, 340, 341
Feedback, 242–44, 247–48, 345
Ferenczi, Sándor, 136, 147–48
Ficino, Marsilio, 104, 313
Figuration, 137–38, 140–42, 345
First Law of Thermodynamics, 261–62
Fludd, Robert, 100, 109–10, 314
Foucault, Michel, 75, 160, 308
Frankfurt School of Social Research, 301, 307
French Academy of Sciences, 111–12
Freud, Sigmund, 147–48, 157–59, 172–73, 256, 323

Index

THE REENCHANTMENT
OF THE WORLD

Designed by Richard E. Rosenbaum.
Composed by The Composing Room of Michigan, Inc.
in 10 point Palatino V.I.P., 2 points leaded,
with display lines in Palatino.
Printed offset by Fairfield Graphics.
Bound by Fairfield Graphics.

Library of Congress Cataloging in Publication Data

Berman, Morris, 1944–
   The reenchantment of the world.

   Includes bibliographical references and index.
   1. Science—Philosophy—History.   2. Science—Social aspects.
3. Reality.   4. Metaphysics.   5. Technology and civilization.
I. Title.
B67.B47   110   81-67178
ISBN 0-8014-1347-8   AACR2